Geographical Places in Transportation

SCIENCES

Geography and Demography, Field Director – Denise Pumain

Infrastructure and Mobility Networks Geography,
Subject Head – Hadrien Commenges and Florent Le Néchet

Geographical Places in Transportation

Coordinated by
Jean-Baptiste Frétigny

iSTE

WILEY

First published 2025 in Great Britain and the United States by ISTE Ltd and John Wiley & Sons, Inc.

ISTE Ltd
27-37 St George's Road
London SW19 4EU
UK

www.iste.co.uk

John Wiley & Sons, Inc.
111 River Street
Hoboken, NJ 07030
USA

www.wiley.com

Library of Congress Control Number: 2025932167

British Library Cataloguing-in-Publication Data
A CIP record for this book is available from the British Library
ISBN 978-1-78945-174-0

ERC code:
SH2 Institutions, Values, Environment and Space
SH2_8 Energy, transportation and mobility
SH2_11 Human, economic and social geography

Contents

Introduction

Jean-Baptiste FRÉTIGNY
PLACES, CY Cergy Paris Université, Cergy-Pontoise, France

I.1. Setting things in motion in and through places

In its significant heterogeneity, transportation forms vast assemblages of actors, infrastructures, tangible and intangible networks, techniques and practices, ensuring the movement of individuals, objects and materials (Mérenne 2014; Cidell 2021; Libourel et al. 2022). While often approached from the angle of the networks brought into play, this book shifts the focus toward the sites where transportation is moored and set in motion, produced and experienced, where it literally takes place.

Let us think about the *Ever Given* ship, which is almost 400 m long and one of the most impressive container ships in the world. Owned by a Japanese group, flying the Panamanian flag, it is operated by a Taiwanese company. Following a sandstorm accompanied by strong winds of up to 74 km/h, it drifted and ran through the Suez canal, blocking all passage to and from on March 23, 2021. Due to this astonishing occurrence, the site of kilometer 151 of the canal became one of the most observed places on the planet. Tension about how long the obstruction would last continued until March 29, when the boat was finally refloated. The site, photos of which are circulating in the media and social networks, stands out in its materiality, its relative but extremely significant narrowness. The place is counted necessarily, in the sense that the dimensions of the waterway are limited in the face of the gigantic size of container ships. The desert environment entails exposure to climatic variations. The diagonally placed container ship is a remarkable event in the supposedly smooth and linear circulation of ships. Resolving the immobility of the *Ever Given*, at the local scale, depends on hundreds of ships, including supertankers, in the canal and on both

Geographical Places in Transportation,
coordinated by Jean-Baptiste FRÉTIGNY.

sides, at the Mediterranean Sea and the Gulf of Suez. Also at play is the end of the freeze on Egypt's crucial source of foreign currency generated by ships in transit and, more broadly, an entire part of the globalized circulation of cash flows that accompany the flow of objects and materials. Supply chain disruption, already under pressure with Covid-19, worries a wide range of institutional actors, with the cost of blocking estimated at nearly $400 million per hour (Vlamis 2021). It is also a source of concern for many people who have purchased computer equipment or other objects on the Internet. Through this very place, at kilometer 151, the *Ever Given* dramatically and, in some respects, unprecedentedly gives transportation major visibility. The human cost of the complex bailout operation has been much less commented on, however. One of the Suez Canal Authority ships sank there and one of the workers on board lost his life (Ankel 2021). Here, the transportation place acquires another dimension, that of an intimate drama and the working conditions that made it possible.

This introduction is specifically devoted to the conceptualization, developed in this book, of how places and transportation interweave. It then points to the importance of broadening the view to rural worlds and to the Global South, in a context where a large part of the existing work concerns urban societies and the Global North. As this volume only partly escapes this bias, the challenge of this development is to contribute to further opening up research avenues. Finally, the presentation of the different chapters, in particular of their critical and humanistic dimension, will help to further elaborate the heuristic value of transportation and its places for the understanding of societies.

Adopting the formula that, in many respects, transportation is "first and foremost a place" (Lombard and Steck 2004), this book draws on the growing literature which, despite not formulating it in these terms with rare exceptions[1], sheds light on the sense of place in transport and, therefore, contributes to understanding its significance for our societies. Transportation places are nodes of physical networks and sites that participate in considerable and evolving technical, travel-related infrastructures. However, this technical complexity is not enough to define them. As places, they are also, according to Doreen Massey's famous theorization, "articulated moments in networks of social relations and understandings, but where a larger proportion of those relations, experiences and understandings are constructed on a far larger scale than what we happen to define for that moment as the place itself" (Massey 1994, p. 154). Through the many activities that take place there, transportation places actively participate in the interdependencies of all kinds

1 The thematic issue edited by Lombard and Steck (2004) represents a particularly noteworthy contribution to their conceptualization.

that form the basis of our daily lives. More than any other places, they imply a form of self-overcoming, of trans-scalarity.

I.2. The wide scope of places related to transportation

The most spectacular overcoming is undoubtedly that of interchange hubs or multimodal platforms, otherwise referred to as "switches" (Lévy and Lussault 2003) or "synapses" (Brunet et al. 1992) to evoke the connections at the very heart of mass transportation systems. Changing from one mode of transportation to another (intermodality) or from one line to another in the same network (intramodality) turns these sites into "movement-places" (Amar 1989) both very connected on a large scale and sometimes very extensive on a finer scale. Let us think of the expansion of underground complexes combining transportation and other activities (retail, offices, housing, etc.) such as the one linking Paris Saint-Lazare train station to Auber, Opéra, Saint-Augustin and Havre-Caumartin suburban railway and metro stations, the subterranean space of La Défense (Grande Arche), or "the inner city" of Montreal. It has earned them the name of "urban mangroves" (Mangin et al. 2016). This distension of these places is paradoxical, since the raison d'être of these sites is precisely their locality, thus presenting a certain unity and specifically limited internal distances to ensure connection. The dimensions of these places are nevertheless sometimes considerable and lead to the insertion of various forms of transportation to reduce the distance: conveyor belts, escalators and elevators, but also in certain places, such as major airports, automated metros, buses or cars to access the plane for first-class passengers, or, in terminals, golf carts for disabled passengers, Segways for the police, bicycles and scooters for certain members of staff, etc. Large or compact, these transportation places involve social activities and interactions – be they to avoid contact or socialize – between people of heterogeneous social and spatial backgrounds.

Smaller transportation sites, such as bus stops, often materialized by a simple sign, are also invested with a major cultural, affective and social life, although little studied. We can think of the "bus-stop culture" of rural teenagers, identified by Moore (2003) in English villages of East Anglia. The social appropriation of a stop placed in a central position in the village, easy to reach and with a grocery store nearby, enables them to meet, drink, eat, or even smoke cannabis, not without some tension with residents who see it as problematic ways of "hanging out". In their use for travel purposes this time, bus stops are also places where – often neglected – emotions can flourish or be appeased, prompted by the risk of being exposed to bad weather, but also to aggressions, on site or on the way. They are also complex performance sites, as illustrated by negotiations to access the bus at peak times.

Studied by Rink (2016) in Cape Town, these negotiations mobilize social codes which are related, quite expectedly, to the proximity to the bus door, but also to the social position of each rider, in terms of power relationships of gender, age, race or disability.

Bus stops are part of a broader set of transportation-related places that can be described as traffic sites, corresponding to all thoroughfares or transportation routes. They are part of another overcoming of transportation places, a dimensional change "from point to line" (Lombard and Steck 2004) – the importance of which is both material and symbolic. Consider, for example, the zero point from which road distances are calculated. They reflect as much the centrality of the chosen location in the road network as the political recognition of its position in the urban order and the geography of power, as shown by Tiananmen Square in Beijing, the top of Capitoline Hill in Rome or the Plaza de Armas in Santiago de Chile.

Traffic sites are a form of planning, regulation and uses of places that facilitates their overcoming, the passage from place to place. Roads have typically been the focus of academic enquiry. This preponderance stems from the dominance of the system of infrastructure, practices and values that define automobility (Beckmann 2001; Flonneau 2008), while it can also be partly explained by the use of roads for other forms of land transportation, particularly active mobilities (walking, cycling, etc.). The study by Baldasseroni and Charansonney (2018) on the information intended for motorists in Paris and Lyon since the interwar period, for example, shows well how the accumulation of information devices over time changes the places traveled through and the relationship with them.

Facilitating access to other sites by setting up signage from place to place was the predominant device from the 1920s to 1940s. In the 1950s to 1970s, interventions in favor of the automobile brought about the installation of traffic lights and their coordination. This was the great period of "green waves", intended to allow motorists, at a particular speed, to meet green lights on the move. This system illustrates the growing concern of elected officials and engineers to improve car traffic flow. The design of the transportation place as a place of transit stumbles on one of its key properties as a place: the fact that internal distances are limited and that congestion is always possible – against the backdrop of a car-dominant appropriation of the place.

The transportation place, from this point of view, is profoundly processual, a "process-place" (Cattan 2012) whose changing situation is decisive in understanding its content. Traffic conditions or the occurrence of works on Paris or Lyon's thoroughfares are a case in point. From the 1980s onwards, real-time traffic information changed the relationship between users and the places they pass through with the reign of the so-called variable message signs placed on the Paris ring road (the *Périphérique*). It is also, and above all, the triumph of on-board information, for which the radio in the car is pivotal. Traffic information is disseminated by the radio data system (RDS), which uses the FM band. The on-board information is then obtained from Internet traffic information tools. In France, they were very popular during major public transport strikes of 1995. This episode reinforced the use of roads but also the consideration of the bicycle and, more broadly, non-car traffic, in the face of congestion. In this digital mediation between users and transportation places, private actors tend to prevail over public actors, who had until then been predominant in the field of information. From the end of the 1990s, in fact, connected objects such as satnavs, then, from 2008 onwards, digital applications on smartphones such as Waze have continued to bring digital platforms to the fore (Baldasseroni and Charansonney 2018). In the event of disruption, digitally optimized routes bypass congested sites and transfer traffic to more residential, narrower roads, shaking up the room usually given to traffic in these more local places and reminding us of how interdependent transportation places are. The study of these places calls into question the sociotechnical devices and the operational challenges but also the power relations between actors that play out through the production and consumption of mobility services.

I.3. Transportation and its places as an immersion in social life

Through its places, transportation introduces a dimensional change from the point to the surface, to the volume and, more broadly, to the social world as a whole. Let us take the "Yes to FRA" communication campaign (*Ja zu FRA*) led by Fraport, the airport authorities of Frankfurt am Main, that is, of Germany's leading airline hub (Figure I.1). It plays with the use of IATA codes[2] (FRA designating Frankfurt Airport), which, by extension, comes to refer to the city as a whole.

The narrative of the airport as a gateway to the urban and national territory aims to strengthen the acceptability of its activity and its territorial and volumetric footprint generated by air traffic. Its legitimacy is indeed questioned by an entrenched environmental conflict. The noise pollution along air corridors has crystallized a long-lasting conflict with residents and environmental activists, which led to the death of two police officers during the construction of the third track in 1987.

2 Defined by the International Air Transport Association, a trade organization which represents the interests of many airlines around the world.

Antagonism continues to fuel weekly protests against the intensity of trafficking in the middle of the departure halls of Terminal 1 (Frétigny 2016). Transportation places such as airports have thus not waited for the rise of climate activism to be targeted by environmental mobilizations.

Figure I.1. *"Because our door to the world must remain wide open". Campaign promoting Frankfurt airport (source: Frétigny (2012)). For a color version of this figure, see www.iste.co.uk/fretigny/transportation.zip*

Transportation places are also often symbolic places of territories, that is to say, in charge of representing and making perceptible the material and ideal relationship of a collective to a given space, deeply imbued with values and ideologies. One of the key papers devoted to this notion (Debarbieux 1995) evokes from the outset the statues embodying the cities of northern France on the Parisian station Gare du Nord's façade and the murals of Paris' Gare de Lyon, which idealize the territories that are served toward the south-east of France. This symbolic dimension is all the more effective in transportation places because it is rarely identified, overshadowed by their highly visible functional nature. These places are nonetheless, more broadly, key vehicles of territorial ideologies, as shown by the signs that the Argentine Ministry of Transport has placed on the country's roads, car parks, train stations and airports to make the claim to the Falkland Islands alive and self-evident, 40 years after the fruitless military confrontation with the United Kingdom to obtain sovereignty (Figure I.2).

Figure I.2. *Pose in front of the slogan "The Falklands unite us" at the station of Retiro, Buenos Aires (source: Frétigny (2022)). For a color version of this figure, see www.iste.co.uk/fretigny/transportation.zip*

This function of a symbolic place can take on more unexpected forms, such as the many vans that display the Ukrainian flag on their license plates or elsewhere. Present in European cities to circulate food, medicine or people to or from Ukraine, in the wake of the Russian invasion of Ukraine in 2022, they embody a material and symbolic bottom-up struggle in favor of its sovereignty (Figure I.3).

Highlighting the underpinnings of transportation and its places in the more encompassing field of political, economic, social, cultural and environmental issues of societies chimes with the perspective developed by mobility studies since the early 2000s. By focusing on mobility in relation to both place and change of place, and as what accompanies, precedes and prolongs travel, this strand contributes greatly to illuminating and contextualizing transportation and its places. One of the most cited programmatic papers on mobilities suggestively devotes an entire section to a transportation place: the airport, presented as a site of high complexity (Sheller and Urry 2006).

Figure I.3. *Van with the colors of Ukraine in Paris (source: Frétigny (2022)). For a color version of this figure, see www.iste.co.uk/fretigny/transportation.zip*

Through the mushrooming of work on transportation places, a broader convergence between mobility and transportation studies is taking place, called for by many authors (Fumey et al. 2009; Shaw and Hesse 2010; Cresswell 2011). Interdisciplinary in nature, it owes much to geographers, but also to historians, pointing to the role of transportation and its places in shaping the social world (Divall and Revill 2009; Flonneau and Guigueno 2009), as well as sociologists and anthropologists (Vincent-Geslin et al. 2019). The rise of rail transportation places like stations, with their large and small clocks, has thus played an essential role in the diffusion of a relationship with time valuing precision and punctuality. While each city had until then had its own local time and people very different time markers, the fulgurant expansion of the railway system has greatly contributed to the standardization of time, with previously locally set time of each aligned for each railway line, and then synchronized at the national level (Harrington 2003). Stations, the practice of which was not self-evident for the many neophyte travelers, were thus the cornerstone of people's socialization to the codes and mobility cultures of modernity (Löfgren 2008).

The convergence between mobility and transportation studies is response to the aspiration for greater reflexivity in transportation geography with regard to its role in the social world (Goetz et al. 2009; Attoh 2014), particularly in relation to the climate emergency and glaring inequalities. To bridge perspectives from transportation and mobility studies can also inspire conversation between the two main axes of transportation geography (Dobruszkes 2012), namely networks, infrastructure and transport services on the one hand, and the consequences of transportation on society as a whole on the other. Transportation places are indeed boundary objects (Star and Griesemer 1989) at the intersection of these fields of study.

I.4. A humanistic and critical approach

By probing transportation through its places, this book adopts a humanistic perspective, in this sense alighting on the humanistic geography developed from the 1970s and 1980s onwards and emphasizing the sense of place and relationship to place, which has contributed to influencing a large part of contemporary social and cultural geographies and, more broadly, of research related to mobilities. Through his attention to the vernacular landscapes of the United States and the role played by the car and the road, the pioneering work of Jackson (1997) illustrates the heuristic value of such an approach. Without giving in to an anthropocentric vision, the challenge is, alongside the functional logics of transportation, to pay heed to the relational nature of transportation and its places, entailing a historical depth, social interactions, power relations, collective imaginations and varied sensory experiences. From this point of view, the popularity of the idea of non-places (Augé 1992), which specifically targeted transportation places, beyond the eminently problematic character of the category, widely attested to by the literature (e.g. Merriman 2004), testifies to the relevance of their study in a humanistic perspective.

The aim of the book is to critically consider these places as key sites through which a considerable part of the challenges of our societies is played out, such as social and spatial justice challenges. The French *Gilets Jaunes* Movement, in the winter of 2018–2019 and in the spring of 2019, in response to the gradual increase in carbon taxation of gasoline prices, is a case in point. Predominantly composed of members of the working and lower middle classes (Collectif d'enquête sur les Gilets jaunes 2019), this movement was not only epitomized by these high-visibility jackets that had just become mandatory in cars. People took over the roundabouts of urban fringes to demonstrate, camp or block them, transforming them into emblematic transportation places refracting issues linked to car dependency in urban peripheries and rural areas. The growing academic interest in transport and mobility (in)justices demonstrates more generally the far-reaching importance of these issues (Verlinghieri and Schwanen 2020).

Transportation places are also very much involved in the challenges for the habitability of the Earth, posed by climate change and the biodiversity crisis, in a context where transportation is responsible for nearly 22% of global CO_2 emissions linked to fossil fuels in 2019[3] (EDGAR 2020), an increase of 78% compared to 1990 emissions. They also play an important role in well-being and physical and mental health, from the perspective of active mobility, in the face of aging and the lack of physical activity related to sedentary lifestyles. Finally, they raise major planning and public policies challenges, linked to congestion, accidents, the place given to activities beyond transportation (be they lucrative or not), the harshness of the working conditions that are linked to these key sites of capitalism and logistics (Cowen 2014), health issues related to epidemics such as Covid-19, etc.

I.5. The multiplicity of fixed or mobile places shaping transportation

Like transportation itself, the places that constitute transportation are particularly heterogeneous. This variety is primarily linked to the growing diversification of transportation modes and networks. Aerial ropeway systems, for example, whether by cable cars or gondolas, are becoming increasingly popular for urban public transport, whereas for a long time they were limited mainly to mountain resorts. Deployment in Latin America, in Medellin in Colombia in 2004, then in Caracas, La Paz, Bogotá and Mexico City, has spread to Portland, Roosevelt Island in New York, and London, Brest in France or Saint-Denis on Réunion Island.

The diversity of these places is also due to the profusion of sites associated with transportation. It is striking when we consider automobility, conceived as the dominant, car-related, transportation system. Its importance is also reflected in the increased number of these places beyond the road: parking lots and parking spaces, garages, service stations, drive throughs for fast-food chains and mass retail, traffic control rooms, etc.

Other transportation systems and services are also major vectors for the proliferation of transportation places. Some new places take shape, such as community bike schools, where people can learn to ride a bike or get back in the saddle. Former shops transformed into warehouses for commercial orders located in urban centers, often referred to as dark stores, are another example. These staples of

3 Base year before the disruptions caused by the Covid-19 pandemic.

logistics chains are used by Uberized workers to deliver orders over their last kilometer, on scooter or by bike.

Many places are closely intertwined with transportation without being identified as such. Think of free zones, products of a neoliberal nature that grant tax and regulatory advantages for imported goods that are then exported. These sites call for rapid circulation and operate in symbiosis with transportation hubs, such as the Jebel Ali Free Zone in Dubai. Linked to the deep-water port of the same name and Al Maktoum airport, it is owned by the DP World port operator. Covering 57 km^2, the equivalent of more than 7,000 football grounds, it is touted as the largest in the world by its owner and is said to be responsible for nearly 24% of the city's GDP (JAFZA 2022). In contrast to these sites of accelerated object circulation, other places associated with stations, ports or airports are based on the logic of restricting individuals' mobilities, detaining migrants in particular, as part of a wider set of holding facilities linked to opressive migration policies (Migreurop 2017).

Transportation places can also be mobile. Much less studied, they constitute an essential line of investigation, particularly relating to power relations and living conditions, evoked in its own way by fiction. An entire section of social, even political, life is played out through these passenger spaces, as shown in Jafar Panahi's movie *Taxi*, released in 2015. This docu-drama was developed after the Iranian authorities banned the director from filming. To circumvent this censorship, he was filmed, as the driver, inside the taxi and in its immediate vicinity. The scenes with the various passengers bring into play the weight of repressive norms in conjugal relationships, in learning at school or in challenging the ban on women's access to sporting events. More broadly, this microcosm allows us to address the question of human rights.

Among these places on the move, those that fall under the umbrella of mobile habitat are major vectors of lifestyles and territorialities. The movie *Nomadland* by Chloé Zhao, released in 2020, tells the story of a woman, played by Frances McDormand, who lost her job in Empire, Nevada, following the 2008 economic crisis. She buys a van and turns it into her domestic space, parked in vast parking lots, swinging from one seasonal workplace gig to another. In particular, she works in Amazon warehouses during peak periods, such as when orders pour in before Christmas. The multinational's system, closely linked to logistics and transportation, is thus based on the exploitation of van dwellers, whose working and living conditions are marked by precariousness, the harshness of daily life and forms of solidarity, documented in the book by journalist Jessica Bruder (2020), of which the movie is an adaptation.

Ordinary mobile places of public transport are also sites of emotion and relationships of domination, particularly through harassment and assaults on female travelers. *Cairo 6, 7, 8*, directed by Mohamed Diab and released in 2010, is inspired by an attack on a young woman in Cairo who had filed a complaint and had to leave the country following a media campaign accusing her of damaging the country's reputation (Florin 2020). It points to the extent of the phenomenon of sexual harassment, in these places and beyond, and the difficulty women face in asserting their rights when they overcome the feeling of shame caused by the attack.

By contrast, the rise of mobile places emblematic of business tycoons and other super-rich, whether on the seas or in the air, participates in logics of power that other social actors are seeking, literally, to deconstruct, through digital means. Super-yachts belonging to Russian oligarchs, present in Mediterranean ports or other tourist sites, have been the subject of intense visual open-source intelligence by activist Internet users to ensure that they do not escape the application of sanctions planned by Western countries in the context of the Russian invasion of Ukraine. These targeted pleasure boats, which often vary in length between 40 and 180 m, had been disconnected by their owners from the maritime automatic identification system (AIS) that facilitates sea navigation, indicating the position of each boat (Ungoed-Thomas and Handy 2022).

The private jet sector is also subject to militant attention in the environmental field. Private jets represent no less than one in 10 take-offs in France according to Eurocontrol data (responsible for the safety of air navigation in Europe) used by the European NGO Transport & Environment (2021, p. 20). The ADS-B system[4], which enables aircraft to regularly broadcast their position by radio, makes their trajectories visible. The X (formerly Twitter) account @Elon Jet, followed by nearly 500,000 subscribers in the summer of 2022, thus reports, in an automated way, each trip and the associated carbon emissions of Elon Musk's private jet. Other social media accounts scrutinize these practices, such as *L'aviondeBernard*[5] on Instagram and X, dedicated to the jet of the luxury company LVMH, headed by Bernard Arnault, one of the richest people in the world. The cartographic data visualizations aim to highlight the extent and singularity of the environmental footprint of these mobile places, and therefore of those who practice them. From the dark store to the private jet, the diversity of places that constitute transportation is refracted by the sheer variety of their protagonists.

---

4 Automatic Dependent Surveillance–Broadcast. Surveillance, that is, air traffic control, is described as dependent because it relies on tools installed onboard aircraft rather than on those exclusive to air traffic controllers themselves.

5 See: www.instagram.com/laviondebernard.

I.7. The heterogeneity of parties involved

The multiplicity of actors involved in transportation and its places is a key factor in their complexity. There is no doubt that public actors can markedly demonstrate their willingness to intervene in these places, especially when they are of strategic importance. Mexican President Enrique Peña Nieto decided in 2014 to transfer Mexico international airport, Benito Juárez, the most important airport in Latin America and largely overtaken by urbanization, 10 km away, to the northeast, in the federal area known as Lake Texcoco. The looming expropriation of residents, land urbanization and job losses in the neighborhoods surrounding Benito Juárez airport sparked local concerns and protests. There were also allegations of corruption focused on the gigantic construction site of the future airport and the contentious allocation of Benito Juárez airport land to real estate developers after its envisioned demolition. With work already underway, Enrique Peña Nieto's successor and political opponent, Andrés Manuel López Obrador, decided in 2018 to cancel the project. In addition to Benito Juárez, he launched a new airport project, 50 km north of Mexico City, on a military base. Named Felipe Ángeles and inaugurated in March 2022, this airport is itself highly controversial (Krauze 2022), showing the extent to which transportation places are instrumental in national and local political games.

The influence of private actors, large or small, is often considerable. Let us think of multinationals: container shipping companies such as MSC, managing very large port terminals, airport and freeway operators such as Vinci or digital platforms of ride-hailing services, as Uber. Other actors operate in less visible sites, participating in the diversification of activities occurring in transportation places, such as The Arch Company (2022) in the United Kingdom. It owns 5,200 railway viaduct arches in England and Wales and looks to rent them out to traders or other operators, such as in Vauxhall in south-west London, famous for its gay nightclubs Paradise and Fire under the station's viaduct. This real estate company presents itself as the largest lessor of small businesses in the country while being owned by another real estate company, Telereal Trillium, and Blackstone Property Partners, a subsidiary of the gigantic American asset manager Blackstone.

These enmeshments between transportation and other activities are manifold. Japanese railway companies have for instance played a major role in the urbanization and the unfolding of Japanese social life around train stations, especially in the suburbs, by developing retail there, and then by building and managing offices nearby (Aveline 2003). The importance of these firms remains unchanged. Tokyo's tallest tower, Tokyo Skytree, is a case in point. Elevator mobility – another transportation place – to enjoy the panoramic view at its top is not only an integral part of the city's tourist circuit, with its light shows, music and acceleration and deceleration effects, just like the shopping mall it overlooks. It is also owned by the Tōbu Railway, a private railway company which built it, with other players, to enhance one of its brownfield sites and boost the number of visitors to Tokyo's lower town through which several of its lines pass.

From a neoliberal perspective, public authorities often seek to involve private actors in the management or financing of transportation places, through privatizations or public-private partnerships. Cable cars crossing the Thames in the docks of London are a striking example of the latter. Their construction, in the run-up to the 2012 Olympic Games, was 60% financed by the airline Emirates and operated by Transport for London, responsible for the city's public transport. For 10 years, from its inauguration to mid-2022, the London Cable Car carried the Emirates name, displaying advertisements for the company's fleet in the cabins and stations. In exchange, Emirates acquired the name of this transportation system, getting the right to write its name literally on the map of London's public transport.

Companies and places of economic activity also show how crucial it is to take a broad approach to transportation and its places. Public policies increasingly emphasize the role of employers in the production of mobilities. Employers can be made responsible for promoting a shift from individual car usage to less emissive practices and incentivized to redesign workplaces to facilitate e-bike charging or encourage pedestrian access. In France, for example, employer mobility plans have been mandatory for urban workplaces of more than a hundred employees since 2018. Such urban planning tool is a de facto recognition of the importance of these sites as locales where transportation takes place.

Beyond institutions, transportation and its places involve individuals and collectives whose analysis is just as fundamental. By conversing with "regular customers, traders, smugglers, and charterers", especially at bus stations, where waiting makes them more available, Choplin et al. (2019) trace "a globalization of the poor", in the wake of sociologist Alain Tarrius (2002). These intense mobilities of merchants, more discreet than the globalization from above of multinationals, cover a multiplicity of transport modes and market places (wholesale, semi-wholesale and retail markets), studied in sub-Saharan Africa and beyond. Relationships at play include border checkpoints, key places where the passage of goods is negotiated – often informally – and invite us to move away from Western-centric approaches focused on immigration. In the ports of Rotterdam or Le Havre, for example, traders come to Europe to ship second-hand goods made in China to Africa and the Middle East, where the quality of these goods is more appreciated than that shipped directly from China.

Attending to riders' viewpoints is also essential, as well as of those who set in motion or maintain transportation places: lorry drivers, drivers of pirogues or tricycles for tourists, boatmen, gardeners looking after the vegetation on major urban streets, etc. Documenting the practices of women working in northern European ports (Lécuyer 2022), for example, by mobilizing the reflections of feminist geography, is an opportunity to elucidate gendering processes at stake in transportation places, an issue of particular importance in the case of ports, given how strongly they are associated with male presence. Alongside such scientific studies, fiction works such as Emmanuel Carrère's movie *Between Two Worlds*, released in 2021 and inspired by the book by journalist Florence Aubenas (2010), contribute to making visible the arduous work and mundane life of women who, as shown in the movie, clean aboard the ferries running between the Normandy port of Ouistreham and Portsmouth.

Paying heed to the experience of transportation places by users is pivotal to challenge the stereotypical representations to which they are exposed. Contrary to the assumptions of decision-makers who focus their attention on saving travel time by creating new public transit lines (Gallez 2015, p. 222), studies show the importance of not considering the time of travel as lost for passengers (Jain and Lyons 2008). Travelers can appropriate the space-time of the transportation place and engage in activities they aspire to such as using Wi-Fi hotspots. More broadly, the challenge is to grasp the multiplicity of the relationships engaged both with transportation places and with distant places by passengers, which are themselves very heterogeneous (Frétigny and Lin 2021). To be sure, attention is also needed to other users of transportation places, such as those in a situation of homelessness, who can be strongly invested in them.

I.8. Shifting the focus beyond the Global North and urban worlds

If the sheer heterogeneity of transportation, its places as well as its actors, is challenging, so is the fact that most studies concern the Global North and urban societies. While I have sought to loosen this dominant prism in the preparation of this volume, this perspective remains largely to be pursued, bearing in mind the necessarily situatedness of every academic study. Developing knowledge freed from colonial logics in the field of transportation is now a growing concern in scholarship (Wood et al. 2020). For example, we need to take apart, as Schwanen (2018) invites us, the risks of casting as a universal model the division of public space in traffic lanes separated by speed: sidewalks, cycle lanes, bus lanes and general traffic lanes. This standardization, for the sake of optimizing speed and limit friction, goes hand in hand with a functionalist mindset of the urban order and (Western) modernity. It bears the risk of legitimizing the eviction of paratransit (also known as "informal transport"): minibuses, motorcycle taxis, tricycles, quadricycles, etc. Paratransit hardly fits in with these predefined spaces, since it is obliterated in their conception, as are alternative uses of the road, such as street vending, made out-of-place. Even though their drivers make their way through traffic jams, paratransit vehicles are often blamed for contributing to congestion. Paratransit, also present in the Global North, especially in tourist sites (Berroir et al. 2016), occupies considerable space in many cities and beyond. But its role is frequently neglected, as shown by its invisibility in official statistics. The same is true for its – potential – connections with other transportation modes, despite being socially innovative, such as carpooling or demand-responsive transportation.

Studies on transportation places in the Global South show their major contribution to the dynamics of societies and to forms of innovation. Collective transportation places in Angola, for example, whether buses, minivans, 4 × 4s or trains, are unexpected sites for the rise of alternative sociopolitical cultures in authoritarian regimes (Neto 2020). These original political arenas, referred to as "moving assemblies", are crystallized through the confrontation of travelers, during their journey, with elements arousing mistrust, such as acts of corruption by police officers, demanding to be paid for each unconventional bag hoisted over a bus, leading to protests. The deplorable state of the transport infrastructure also acts as a trigger, as well as criticizing the exploitation of natural resources by Chinese companies, authorized by the Angolan president, at the sight of their trucks. Passing through the massacre sites of the Angolan civil war, whose memory is eminently political, also fuels the discussion. This emergent politicization of citizens through transportation places is difficult to control for local authorities, because of the number of people using these places and their constant movement.

Rural transportation and its places also receive limited attention. To be sure, transportation infrastructure is less present in rural areas, raising major questions of spatial justice. While the population is less dense, the places to be reached are more distant. A study in four counties in Wales, for example, showed that 71% of respondents had a bus stop accessible within walking distance. However, 69% of them considered that the service provided was "poor" or "very poor quality", citing mainly the low frequency of buses, contributing to explain why only 29% of them used buses (Milbourne and Kitchen 2014). Bus frequencies have worsened due to austerity policies after the 2008 economic crisis, exacerbating the issues of car dependency.

Does the lack of infrastructure in rural areas lead to relativizing transportation and its places or, on the contrary, to their increasing visibility in everyday life and in people's concerns? Porter's (2016) study on Tanzanian villages without road access in the district of Kibahia suggests the latter. It points to the major role played by motorcycle taxis (*boda-boda*) from the end of the 2010s, with the irruption of low-cost models from China, against a background of no less intensive practices of walking and riding bicycle taxis. The materiality of the road, more or less passable, captures people's attention, raising considerations ranging from its practicability outside of the dry season to the presence of potholes, to the point of appearing as the major concern about travel for women developing their own economic activity in rural areas of Nigeria (Seedhouse et al. 2016).

Figure I.4. *Near the school of Kaupanger, a Norwegian village: "arrival zone [of children at school]", "heart zone" (source: Frétigny (2021)). For a color version of this figure, see www.iste.co.uk/fretigny/transportation.zip*

An equally important topic to address is the innovation at work in rural areas. Let us think of NGOs such as the Soft Vehicle Innovation association (In'VD 2022), located in a highland region in south-central France, in Castelnau-Pégayrols, near Millau. Its members leverage the attraction of vehicle-related innovations to initiate a broader transformation of road mobilities. They develop prototypes and acquire hybrid vehicles, intermediate between the bicycle and the car, such as tricycles or quadricycles with electric assistance. They are also called light electric vehicles, or in this NGO's words, "soft vehicles". Designed for multiple passengers and to transport objects, they are strongly reminiscent of paratransit vehicles. Other actors seek to shape mobilities by redesigning places linked to transportation such as schools in Norwegian villages where a car-free "core zone" is defined, forcing drivers to park further away to encourage walking and limit congestion, local pollution and accidents affecting children (Figure I.4).

Riders also creatively self-organize their transportation as shown by mundane practices in the outskirts of urban areas. Studying commuting in the rural-urban fringes of Grenoble and Aix-en-Provence, in the southeast of France, Pradel et al. (2015) identify multiple informal and inventive collective assemblages: parking lots acting as carpooling areas, storage of bicycles in the trunk during carpooling to use them on arrival at the cycle path, etc. There is no doubt that rural areas and the Global South are essential areas for expanding our understanding of transportation and its places.

I.9. Structure of the book

The book's chapters provide a humanistic and critical analysis of transportation and its places by probing their diversity. Unlike many works, each chapter addresses transversally the different modes of transportation and associated places, while aiming to diversify the national contexts of investigation.

By drawing on the perspective of geohistory, the first chapter opens a passage between transport networks, analyzed through the lens of their deployment and mutations, and the related places, which are explored in more details in the following chapters of the book. Anne Bretagnolle and Christophe Mimeur highlight the extent to which networks foster the emergence of highly accessible places, working as major transportation places and developed in close connection with the city they serve. The inclusion of transportation and its places in long-term timelines and on different time scales is at the heart of the humanistic approach of this chapter, prompting us to nuance the seemingly unprecedented nature of contemporary changes affecting them. Its critical dimension lies in its exploration of the logics of domination that favor certain places and exclude others. The authors thus pinpoint transportation places that have been short-circuited by the restructuration of networks. The accentuated hierarchy of transportation places not only leads to the primacy of most central places in the networks, greatly benefitting from the processes of accumulation and concentration of populations and resources, but also to the marginalization of previously prominent places.

While this first chapter approaches the situation and accessibility of transportation places, Chapter 2 examines the role of interface that these places play, insofar as they are part of larger assemblages. These sites are indeed the physical and digital gateways to territories, highly interdependent and requiring the alignment of multiple actors, which are at the heart of their study. Nadine Cattan and I question these gateways as one of the most vivid spatial expressions of globalization "in the making". Authors make sense of the great diversity of gateways by conceiving them in four major ways. They are identified as site gateways – such as stations or (air)ports – or corridor gateways – such as large international straits – or even as territorial gateways – such as transitional immigrant neighborhoods[6] of metropolises, marked by the intense circulation of objects, people and information on an international scale. But through the relational lens of people themselves, subject gateways can also be envisioned, conceived as individuals who by themselves, in situ, ensure international connections, exemplified by Amazon delivery drivers who embody, hastily and at the doorstep of homes, the ultimate gateways to globalized transportation. The humanistic and critical perspective of this chapter stems from the attention paid to the actors of these gateways, whether regarding the challenging – or even elusive – governance of these gateways, the "informal alongside the official" or the role of the subject gateways and their bodies. It also lies in the discussion of the ideological dimension of gateways, highly politicized, in the context of geopolitical games between states, with their social, humanitarian (from a migratory perspective) and environmental consequences also being highly contested across multiple spatial scales.

Chapter 3 delves into transportation places to illuminate the economic rationales underlying their development and operational management. Juliette Maulat and Mathilde Pedro highlight the growing enrolment of private actors and funding in their business model. Shaped by complex neo-liberal processes that affect urban infrastructures more broadly, these sites are increasingly considered as assets whose value needs to be enhanced from a commercial, advertising and real estate point of view. Through the study of railway stations and airports, the authors explore the

6 These are neighborhoods in urban centers of diasporic communities, where migrants tend to settle upon arriving in the city, before moving elsewhere. These neighborhoods host a large variety of activities (commercial, religious, festive, etc.) for the diaspora on the scale of the metropolis and beyond.

processes of the growing diversification of funding sources for transportation places. Market liberalization, the success of new public management policies and the privatization of transportation operators are major drivers to these changes. Additionally, the international circulation of urban planning models, such as the real estate development of railway stations in Japan, inspired by the United States and now spreading to Europe, plays a significant role. The critical and humanistic content of the chapter lies in the deconstruction of these co-dependent dynamics, as well as in questioning the scope of the resulting changes in the business model. While the diversification of the activities engaged by transportation infrastructure managers leads in some cases to developing social housing on their land, it tends most often to favor upper-class travelers and consumers, by building housing and offices to accommodate them. These class-based policies pursued for the sake of financialization affect the experience that users of railway stations and airports have. They also lead to strong discrepancies and tensions with the expectations of local actors, confronted with unwanted rising prices, commercial competition and urban densification. In addition, Juliette Maulat and Mathilde Pedro expose the unstable nature of this diversification funding, which can undermine the proper functioning of transportation places. They also insist on the inequalities that these changes create between transportation places, and consequently between (urban and rural) regions, fostering a two-tiered network, where major transportation places generate the expected resources, unlike intermediate or smaller places, whose financing and their experience by users are affected by the lack of care, employees or adapted waiting spaces.

It is also as key sites of capitalism that transportation places are envisioned in Chapter 4, scrutinizing those specialized in the mobility of objects, from the standpoint of logistics. Nicolas Raimbault explores the very male-dominated world of delivery drivers and warehouse workers, such as forklift drivers. Logistics operatives represent a large and increasing segment of the contemporary working class and of vehicles on the road. Tellingly, in Paris, with the rise of e-commerce, a quarter of all motor vehicles in circulation are dedicated to deliveries. Drawing on examples from Europe and North America, the author lays out a geography of logistics workers and their places, showing how much it reshapes the contemporary popular centralities of urban regions. Logistics parks, and the associated employment, are developing far beyond traditional working-class settlements in the industrial suburbs, typically on the edge of urban fringes. Situated far from workers' home, they involve long daily commutes. By examining delivery services, especially concentrated in urban centers, Raimbault shows that employers demand such a fast pace that complying with the traffic code becomes challenging. These delivery rhythms threaten road safety for both drivers and other road users. They are all the more arduous due to increasing restrictions on the access of motorized vehicles to urban centers. The humanistic and critical perspective developed in this chapter highlights the precariousness and the strenuousness of their daily working conditions, as shown by the intensified work in warehouses, leading to a rise in accidents. Challenges related to the appropriation of workplaces also emerge, due to a lack of social and rest areas, whether in logistical zones in the urban fringes or in the dark kitchens and dark stores of urban centers. Against a backdrop of very limited union presence, the author also highlights the emerging politicization of these places, to the point of becoming pioneering sites for (inter)professional social mobilizations and thus achieving unprecedented visibility.

The variety of activities deployed at transportation places, discussed in Chapters 3 (with regard to commercial activities) and 4 (regarding food and rest areas at the workplace), lies at the heart of Chapter 5, which addresses public spaces. Antoine Fleury and I show that the relationship between public spaces and transportation is one of entrenched interdependence and ambivalence. They are both closely intertwined and yet in tension, since public spaces emerge when transportation operations allow room for other activities. The "alternative approach of transportation" adopted in this chapter consists of considering the assemblage of transportation as part of a wider space sharing. The humanistic dimension of the chapter is perceptible in the grasp of (im)mobile transportation places as sites where social life unfolds in the mere heterogeneity of its forms. It is also at play by attending to transportation sites whose recognition as public spaces is either minimal or non-existent: parking lots and parking spaces, business and industrial parks, country tracks, etc. Recognizing the skills and professional cultures of institutional actors and, more broadly, imaginaries as determining factors in the planning of transportation-related public spaces is also part of a humanistic stance. The critical character of the analysis emerges in deconstructing the often taken-for-granted "reclaiming" of streets by public transportation, pedestrians, cyclists and drivers of personal transportation vehicles such as light (non-) electric vehicles (scooters, hoverboards, etc.). The analysis of the fair sharing of space also attests to a critical stance by questioning the premium placed on transportation and emphasizing the need for social justice, as exemplified by the creation of inclusive railway station forecourts for all users. The chapter also addresses the challenge of rethinking how space is shared with the non-human, in particular with plants, as seen in the removal of pavement from traffic sites.

This place sharing is then discussed in Chapter 6 in the context of active mobilities (especially walking and cycling) in cities of the Global North. Jean-Paul Hubert, Kaduna-Eve Demailly and Clément Dusong situate their reflection in the wake of the threefold sequence popularized by Newman and Kenworthy (1996), describing the transition from a pedestrian-dominated city (walking city), predominant until the mid-19th century, to a transit-oriented city (transit city), largely shaped by trains until World War II, and subsequently to an automobile city, particularly visible in the United States and Australia. Focusing on the potential transition to a more active mobilities-friendly urban model, the authors explores the role of public policies in facilitating – or hindering – such a shift. They point to the importance of overcoming the fragmentation of traffic routes dedicated to cycling and walking. By designing an infrastructure specific to walking, involving a wide variety of places belonging to heterogeneous actors, the authors (re)legitimize walking as a form of transportation while considering the "pedestrian presence" beyond its utilitarian function, including stillness or recreational movements, such as the children's game of hopscotch. This humanistic approach is accompanied by a call for urban living that engages with active mobilities. The chapter further presents a critical examination of the objectives and tools used in urban planning. The authors put forward the discrepancy between the importance of walking, versus cycling, in individuals' mobile practices and the reverse priority that public action gives to these two forms of transportation. One explanatory factor discussed in the chapter is the specific activism in favor of walking, which, in contrast to the highly sectoral approach prevalent in public policies, encompasses more transversal concerns: childhood, disability, aging, accident prevention, health, heritage, etc. Observing the contrast between Dutch family-oriented cycling practices, using large cargo bikes on the country's wide cycling paths, and Japanese cycling habits favoring oblong bicycles, able to weave onto often undifferentiated paths, the post-car city model and its places appear, at this stage, to be far less uniform in the Global North than previous ones.

Chapter 7 approaches the sharing of transportation places from the perspective of sensitivity and ambiances, which are receiving increasing attention in the literature. Damien Masson shows how these environments are eminently imbued with the circulation of affects, with varying intensities of "transport", to echo the literary and historical meaning of the term referring to acute feelings. After outlining the semantic, theoretical and methodological field related to ambiances, atmospheres and sensory experiences, and linking them to mobility studies, the author approaches these objects from the users' perspective, inviting us to think of them as literally embodied, made of flesh and bones, as well as from the perspective of public transportation operators. Deconstructing the idea that each form of transportation has its own distinct, stable and ever-true ambiance, he instead emphasizes its variability, shaped by the evolving sensory environment, practices and interactions, the sensory echo of lived situations. The micropolitical aspect of ambiances clearly emerges, as they are tied to the circulation of affects creating ambiances of discomfort and anxiety, unfavorable to the contestation of inequalities, or conversely, ambiances of joy or connivance, for example, in highly controlled environments – in terms of temperature, humidity or neutralization of odors and sounds deemed undesirable. The critical and humanistic approach of the chapter is also perceptible in its exploration of the complexities faced by transportation operators in shaping atmospheres. Such interventions are constrained by users' unpredictable appropriations of spaces, leading to the emergence of unforeseen ambiances. The limits to the operators' ambitions of shaping ambiances abound. For instance, designing a subway station to promote "natural" light, favoring the visible register in its design, inevitably affects other sensory registers, especially sound, with poor acoustic performances when the metro arrives or announcements are made. Regarding safety, Masson highlights the contradictions of operators, contributing to the image of a dehumanized metro by limiting staff presence while aiming to foster a sense of safety. He also underscores the ethical and political limitations of a wide range of actions taken in the name of this feeling.

While Chapter 7 explores what permeates the experience of transportation places (ambiances, affects and sensory environment), Chapter 8 addresses the role of urban and social animation endorsed by artisanal transportation. Starting from the observation that artisanal transportation now holds a crucial position in the cities of the Global South, Solène Baffi shows that the places formed or associated with these minibuses, three-wheelers or motorcycle taxis form an insightful terrain for understanding developing metropolises. Through the case of African cities, the author presents the places of artisanal transportation as spatial, temporal and normative in-betweens of urban societies. Employing freshly arrived migrants, these sites foster diversified activities, particularly trade, and an intense social life. They are pioneers in providing transportation to urban fringe areas, working as vectors for the development of rapidly expanding cities. The critical scope of the chapter lies in the analysis of power relations, manifest in conflicts over land use of these places. However, power struggles also arise from the legal and fiscal ambiguities surrounding this sector and its places, facilitating violent confrontations between actors, linked to ethnic and political tensions, or even criminal activities. The ambiguity of public actors transpires through the frequent exclusion of artisanal transportation and its workers from planning policies. These assignments to informality as well as processes of relegation and erasure may sometimes give way to more complex forms of regulation, including concerted or imposed formalization and standardization, or even a more deliberate hybridization with other forms of public transportation[7]. Baffi highlights the ambivalence of this hybridization with the development of digital platform transportation services at the expense of artisanal transportation operators while

7 Public transportation is traditionally contrasted with private transportation and collective transportation with individual transportation. Transportation is considered public if it is organized for others as part of a commercial service, as opposed to private transportation. Public transportation is often collective (trains, buses, commercial flights, etc.). Taxis and ride-hailing services also fall under public transportation. However, unless shared with other passengers traveling in the same direction (such as with UberX Share), they are seen as part of individual transportation.

borrowing many of their practices. The humanistic dimension of the discussion is tangible in the description of the vernacular knowledge at play in the mobile cultures of artisanal transportation, particularly through the highly scripted gestures used to communicate between the street and the vehicle. It is also striking in the analysis of the unique contributions these vehicles make to the urban landscape and to the affirmation of the drivers' identity.

The strength of the power relations affecting artisanal transportation resonates in the investigation of mobile segregation to which Chapter 9 is devoted. Stressing that this issue has received much less attention than residential segregation, in this chapter I probe the range of situations where transportation separates. Segregation in and through mobility has been rarely studied in a comprehensive way, despite its significance to understand how transportation places shape and reproduce hierarchies and social order. The critical approach adopted, building on the frameworks of mobility and transportation justices, highlights the social hierarchy between transportation places, which aligns to some extent with the theory of splintering urbanism, cognizant of socially stratified networks (Graham and Marvin 2001). However, this perspective is nuanced by foregrounding the multiplicity of social relations at work, which are not solely related to class, and by pointing out the importance of segregation within transportation places themselves. The chapter's humanistic framing can be seen in the heed paid to transport places that are often overlooked or rendered invisible, ranging from airport detention centers to precarious homemaking on the Indian street edges. It is also perceptible in the analysis of the bodily and emotional experiences of transportation workers, be they sailors on flag-of-convenience ships or flight attendants, deeply affecting the connection to their (mobile) workplace.

Reflecting the long-term view of the first chapter, Chapter 10 takes us down memory lane, positioning transportation and its places in light of heritage-making. Addressing the tension between the functional logic of transportation and the symbolic processes of heritage, Pierre-Louis Ballot focuses his study on transportation places that are still operational. He explores these places by transportation modes, which, along the related mobile cultures, influence their heritage-making or heritagization. The author examines the iconic UNESCO World Heritage designations, related to transportation places, and explores what motivates them, particularly in the cases of the port of Liverpool and the Chhatrapati Shivaji railway station in Mumbai. He also chisels away at heritage roads and reveals the depth of interrelations between the thoroughfare itself and the territory it serves – to which it is identified – as shown in particular by the national road 7 (NR 7), connecting Paris to the Côte d'Azur. Motivated by a desire to nourish memories and promote tourism, local associations and governments restore road signs, mural advertisements and establish museums. They also organize various events, notably festive commemorations of RN 7's notorious traffic jams during the vacation season in the 1950s–1960s. The following examination of airports illustrates well the humanistic and critical mindset developed in this chapter, highlighting their challenging progression toward recognition as legitimate heritage places. This is partly due to the association of airports with values of modernity in dominant representations. It can also be explained by the influence of economic concerns on identity portrayal, in the form of what the author describes as "heritage-making of the present". However, even these places, with their increasingly inherited infrastructures, do not completely escape from some level of memorial recognition.

Altogether, why transportation places matter lies in their status as "weak places" (Lévy 2000) rather than "non-places". "Weak ties" (Granovetter 1973), as opposed to strong ties (maintained with significant others), or, in another sense, short-duration or infrequent connections, are often predominant in these sites. Yet, "what takes place there can be profoundly impactful" (Lévy 2000, p. 169). Position conferred to minorities, to the living and to marginal spaces, access to employment, working conditions, (geo)political mobilizations and tensions, contrasting circulations of affects, market-driven and security processes, the expansion of digital capitalism, experiences of co-presence in shared spaces, imaginaries of proximity and distance... Transportation places are undoubtedly elastic sites, stretched toward other places, while at the same time assuming a multifunctionality, a symbolism and a habitability of their own. As such, they are the product of multiple dynamics. The dominant representations, which this book seeks to challenge, largely underestimate their social importance and heuristic character. Taken seriously, they help us grasp considerable facets of our lives.

I.10. References

Amar, G. (1989). Lieu-mouvement : les enjeux de l'évolution des stations de métro. Research Report, RATP, Paris.

Ankel, S. (2021). One person reportedly died while helping free the Ever Given ship, the Suez Canal Authority says. *Business Insider* [Online]. Available at: https://www.businessinsider.com/suez-canal-authority-person-died-during-ever-given-rescue-mission-2021-6 [Accessed 23 August 2022].

Attoh, K. (2014). Imagining a "cultural turn" in transportation geography. *Journal of Cultural Geography*, 31(2), 141–151.

Aubenas, F. (2010). *Le quai de Ouistreham*. Olivier, Paris.

Augé, M. (1992). *Non-lieux*. Le Seuil, Paris.

Aveline, N. (2003). *La ville et le rail au Japon : l'expansion des groupes ferroviaires privés à Tokyo et Ôsaka*. CNRS, Paris.

Baldasseroni, L. Charansonney L. (2018). Gouverner la voirie urbaine par l'information de l'automobiliste. Une comparaison Lyon-Paris, des années 1920 à nos jours. *Flux*, 113, 24-40.

Beckmann, J. (2001). Automobility–A social problem and theoretical concept. *Environment and Planning D: Society and Space*, 19(5), 593–607.

Berroir, S., Bouloc, C., Cattan, N., Fleury, A., Frétigny, J.-B., Lassaube, U. (2016). La tour Eiffel dans la poche. Informalité et pouvoir dans la construction de la métropole touristique. *L'Espace Politique*, 29(2). doi: 10.4000/espacepolitique.3876.

Birtchnell, T., Savitzky, S., Urry, J. (eds) (2015). *Cargomobilities: Moving Materials in a Global Age*. Routledge, New York.

Bruder, J. (2020 [2017]). *Nomadland.* J'ai lu, Paris.

Brunet, R., Ferras, R., Théry, H. (1992). *Les mots de la géographie : dictionnaire critique*. Reclus, Montpellier.

Cattan, N. (2012). Trans-territoire. Repenser le lieu par les pratiques spatiales de populations en position de minorité. *L'Information géographique*, 76(2), 57–71.

Chakrabarty, D. (2000). *Provincializing Europe: Postcolonial Thought and Historical Difference*. Princeton University Press, Princeton.

Choplin, A., Pliez, O., Steck, J.-F. (2019). Sur la route de La mondialisation des pauvres : entretien avec Armelle Choplin et Olivier Pliez, par Jean-Fabien Steck. *Echo Géo*, 49. doi: 10.4000/echogeo.18161.

Cidell, J. (2021). *An Introduction to Transportation Geography: Transport, Mobility, and Place*. Rowman & Littlefield, Lanham.

Collectif d'enquête sur les Gilets jaunes (2019). Enquêter *in situ* par questionnaire sur une mobilisation. Une étude sur les gilets jaunes. *Revue française de science politique*, 69(5/6), 869–892.

Cowen, D. (2014). *The Deadly Life of Logistics: Mapping Violence in Global Trade*. University of Minnesota Press, Minneapolis.

Cresswell, T. (2011). Mobilities I: Catching up. *Progress in Human Geography*, 35(4), 550–558.

Debarbieux, B. (1995). Le lieu, le territoire et trois figures de rhétorique. *Espace géographique*, 24(2), 97–112.

Divall, C. and Revill, G. (2009). Les cultures du transport : représentation, pratique et tech-nologie. In *De l'histoire des transports à l'histoire de la mobilité ? État des lieux, enjeux et perspectives de recherche*, Flonneau, M. and Guigueno, V. (eds). Presses Universitaires de Rennes, Rennes.

Dobruszkes, F. (2012). Stimulating or frustrating research ? Transport geography and (un)available data. *Belgeo*, (1/2). doi: 10.4000/belgeo.7082.

EDGAR (2020). Fossil CO_2 emissions of all world countries [Online]. Available at: https://edgar.jrc.ec.europa.eu/report_2020 [Accessed 13 June 2022].

Flonneau, M. (2008). *Les cultures du volant : essai sur les mondes de l'automobilisme, XXe-XXIe siècle*. Autrement, Paris.

Flonneau, M. and Guigueno, V. (eds) (2009). *De l'histoire des transports à l'histoire de la mobilité ? État des lieux, enjeux et perspectives de recherche*. Presses Universitaires de Rennes, Rennes.

Florin, B. (2020). *Les femmes du bus 678* et le harcèlement sexuel en Égypte. In *Abécédaire de la ville au Maghreb et au Moyen-Orient*, Florin, B., Madoeuf, A., Sanmartin, O., Stadnicki, R., Troin, F. (eds). Presses universitaires François Rabelais, Tours.

Frétigny, J.-B. (2016). La conflictualité de la ville en mouvement au regard d'espaces publics métropolitains émergents : les aéroports. In *La ville conflictuelle. Oppositions – Tensions – Négociations*, Desponds, D. and Auclair, E. (eds). Le Manuscrit, Paris.

Frétigny, J.-B. and Lin, W. (2021). Changing geographies of the passenger: Heterogeneous subjects on the move. *Journal of Transport Geography*, 92, 103006. doi: 10.1016/j.jtrangeo.2021.103006.

Fumey, G., Varlet, J., Zembri, P. (eds) (2009). *Mobilités contemporaines. Approches géoculturelles des transports*. Ellipses, Paris.

Gallez, C. (2015). La mobilité quotidienne en politique. Des manières de voir et d'agir. HDR Thesis, Université Paris-Est Marne-la-Vallée, Champs-sur-Marne [Online]. Available at: https://halshs.archives-ouvertes.fr/tel-01261303.

Gallez, C. and Kaufmann, V. (2009). Aux racines de la mobilité en sciences sociales. In *De l'histoire des transports à l'histoire de la mobilité ? État des lieux, enjeux et perspectives de recherche*, Flonneau, M. and Guigueno, V. (eds). Presses Universitaires de Rennes, Rennes.

Goetz, A.R., Vowles, T.M., Tierney, S. (2009). Bridging the qualitative–quantitative divide in transport geography. *The Professional Geographer*, 61(3), 323–335.

Granovetter, M. (1973). The strength of weak ties. *American Journal of Sociology*, 78(6), 1360-1380.

Harrington, R. (2003). Trains, technology and time-travellers: How the Victorians re-invented time [Online]. Available at: http://web.archive.org/web/20080828054933/http://www.greycat.org/papers/timetrav.html [Accessed 1 August 2022].

In'VD (2022). Innovation Véhicules Doux [Online]. Available at: https://www.facebook.com/invd.innovationvehiculesdoux.1 [Accessed 5 August 2022].

Jackson, J.B. (1997). *Landscape in Sight: Looking at America*. Yale University Press, New Haven.

JAFZA (2022). Jebel Ali free zone [Online]. Available at: https://www.jafza.ae/ [Accessed 1 August 2022].

Jain, J. and Lyons, G. (2008). The gift of travel time. *Journal of Transport Geography*, 16(2), 81–89.

Krauze, L. (2022). Mexico City's distant, underwhelming new airport already faces headwinds. *Washington Post* [Online]. Available at: https://www.washingtonpost.com/opinions/2022/03/23/mexico-city-airport-aifa-lopez-obrador/ [Accessed 1 August 2022].

Lécuyer, M. (2022). La féminisation des ports à l'arrêt ? Les pratiques des employeurs et leurs conséquences dans deux ports nord-européens. *Les Cahiers du MIMMOC*, 28. doi: 10.4000/mimmoc.10878.

Lévy, J. (2000). Les nouveaux espaces de la mobilité. In *Les territoires de la mobilité*, Bonnet, M. and Desjeux, D. (eds). Presses Universitaires de France, Paris.

Lévy, J. and Lussault, M. (eds) (2003). *Dictionnaire de la géographie et de l'espace des sociétés*. Belin, Paris.

Libourel, E., Schorung, M., Zembri, P. (2022). *La géographie des transports. Territoires, échelles, acteurs*. Armand Colin, Paris.

Löfgren, O. (2008). Motion and emotion: Learning to be a railway traveller. *Mobilities*, 3(3), 331–351.

Lombard, J. and Steck, B. (2004). Quand le transport est d'abord un lieu ! *Autrepart*, 32(4), 3–19.

Mangin, D., Girodo, M., Seura architectes (2016). *Mangroves urbaines : du métro à la ville : Paris, Montréal, Singapour*. Carré, Paris.

Massey, D. (1994). *Space, Place and Gender*. Polity Press, Cambridge.

Mérenne, É. (2014). *Géographie des transports*. Presses Universitaires de Rennes, Rennes.

Merriman, P. (2004) Driving places Marc Augé, non-places, and the geographies of England's M1 motorway. *Theory, Culture & Society*, 21(4/5), 145–167.

Migreurop (2017) *Atlas des migrants en Europe. Approches critiques des politiques migratoires*. Armand Colin, Paris.

Milbourne, P. and Kitchen, L. (2014). Rural mobilities: Connecting movement and fixity in rural places. *Journal of Rural Studies*, 34, 326–336.

Moore, S. (2003). Hanging about: The importance of bus-stop culture. *Youth and Policy*, 82, 47–59.

Neto, P.F. (2020). Moving assemblies: Socio-political mobilization in Angola's collective transport. *Cultural Studies*, 34(1), 95–121.

Newman, P.W. and Kenworthy, J.R. (1996). The land use–transport connection: An overview. *Land Use Policy*, 13(1), 1–22.

Porter, G. (2016). Mobilities in rural Africa: New connections, new challenges. *Annals of the American Association of Geographers*, 106(2), 434–441.

Pradel, B., Chardonnel, S., Cailly, L., Fourny, M.-C. (2015). Les routines de déplacement dans les espaces périurbains : les dimensions collectives des agencements quotidiens. *Espace populations sociétés*, 1–2. doi: 10.4000/eps.5961.

Rink, B. (2016). Race and the micropolitics of mobility: Mobile autoethnography on a South African bus service. *Transfers*, 6(1), 62–79.

Schwanen, T. (2018). Towards decolonised knowledge about transport. *Palgrave Communications*, 4(1). doi: 10.1057/s41599-018-0130-8.

Seedhouse, A., Johnson, R., Newbery, R. (2016). Potholes and pitfalls: The impact of rural transport on female entrepreneurs in Nigeria. *Journal of Transport Geography*, 54, 140–147.

Shaw, J. and Hesse, M. (2010). Transport, geography and the "new" mobilities. *Transactions of the Institute of British Geographers*, 35(3), 305–312.

Sheller, M. and Urry, J. (2006). The new mobilities paradigm. *Environment and Planning A*, 38(2), 207–226.

Star, S.L. and Griesemer, J.R. (1989). Institutional ecology, translations and boundary objects: Amateurs and professionals in Berkeley's Museum of Vertebrate Zoology, 1907–39. *Social Studies of Science*, 19(3), 387–420.

Tarrius, A. (2002). *La mondialisation par le bas. Les nouveaux nomades de l'économie souterraine*. Balland, Paris.

The Arch Company (2022). About the Arch Company [Online]. Available at: https://www.thearchco.com/about-us/ [Accessed 3 August 2022].

Transport & Environment (2021). Private jets: Can the super rich supercharge zero-emission aviation? Report, Transport & Environment [Online]. Available at: https://www.transportenvironment.org/discover/private-jets-can-the-super-rich-supercharge-zero-emission-aviation/ [Accessed 2 August 2022].

Ungoed-Thomas, J. and Handy, G. (2022). Revealed: Russia-linked superyachts "going dark" to avoid sanctions threat. *The Observer* [Online]. Available at: https://www.theguardian.com/world/2022/may/28/revealed-russia-linked-superyachts-going-dark-to-avoid-tracking-systems [Accessed 2 August 2022].

Verlinghieri, E. and Schwanen, T. (2020). Transport and mobility justice: Evolving discussions. *Journal of Transport Geography*, 87, 102798.

Vincent-Geslin, S., Oppenchaim, N., Belton-Chevallier, L. (2019). *Manuel de sociologie des mobilités géographiques*. Presses universitaires François Rabelais, Tours.

Vlamis, K. (2021). The giant ship stuck in the Suez canal is costing the global economy an estimated $400 Million per hour. *Insider* [Online]. Available at: https://www.businessinsider.com/boat-stuck-suez-canal-costing-estimated-400-million-per-hour-2021-3 [Accessed 18 August 2022].

Wood, A., Kębłowski, W., Tuvikene, T. (2020). Decolonial approaches to urban transport geographies: Introduction to the special issue. *Journal of Transport Geography*, 88, 102811. doi: 10.1016/j.jtrangeo.2020.102811.

1

Infrastructure Networks and Major Transportation Places in the World: A Geohistorical Perspective

Anne BRETAGNOLLE[1] and Christophe MIMEUR[2]

[1]Géographie-cités, Université Paris 1 Panthéon-Sorbonne, Paris and Aubervilliers, France

[2]MATRIS, CY Cergy Paris Université, Cergy-Pontoise, France

1.1. Introduction

At the start of the book, this chapter sheds light on the network dynamics that influence the creation and evolution of transportation sites, helping to put them in perspective. Transportation infrastructure networks are made up of a succession of lines (materialized or not in space) and access points, generally characterized by a high level of attractiveness due to their ability to articulate different scales: local, regional and even global. Because what circulates on these networks – information, populations, products and wealth – benefits from two major assets: speed and increased load capacity compared to off-grid traffic. Entry and exit points, including those which are highly accessible due to a crossroads or interface position, can thus be described as major nodes. A quick overview of the history of cities shows that all major cities, capitals or metropolises, were, at the height of their influence, particularly well-served by the most efficient transportation systems of their time. What role did transportation infrastructure networks play in these exceptional stories? Some will argue that the interactions between transportation and cities

Geographical Places in Transportation,
coordinated by Jean-Baptiste FRÉTIGNY. © ISTE Ltd 2025.

endlessly replay the inextricable paradox of the chicken and the egg: transportation adapts to urban dynamics and in turn is shaped by this same dynamic, with causes and effects being indisputable. While this is generally true, it is nevertheless possible, with the benefit of historical hindsight and the geohistorical approach, to present the major transformations in the field of transportation, whether in terms of technical or organizational innovations that have had an impact on the geography of these major nodes. The processes that then appear, bypassing ancient stages that have become useless, the hierarchical selection of the best-served nodes, regional shifts, or even collapses due to major changes in traffic patterns, are definitely a part of what is referred to as the structuring effects of transportation on territories (Offner 1993; Bretagnolle 2014; Mimeur and Thévenin 2020).

This chapter thus proposes a diachronic approach to transportation infrastructure networks, tackled through the prism of modes of supply production and their technical structure. The chapter is divided into three main periods. The first spanned several millennia and is characterized by the slow and timeless rhythms of human, animal and natural energy. Beneath this apparent inertia, there was fierce competition between the world's major cities to capture major trade routes and secure military roads. Other issues concerned the organization of traffic within the administered territories. The second period spanned the 19th century and part of the 20th century. Over the course of this century and a half, the invention of mechanical energy revolutionized modes of movement, production and even representations of space. Cumulative processes were set in motion, with their associated gigantism and concentration on scales hitherto unequalled, and solutions had to be found to stem the flows and house the populations that flocked to the world's metropolises. The third period began in the aftermath of the Second World War and covers the current phase of economic globalization. Structuring transportation networks are largely intangible: container shipping lines, freight and traveler airlines, internet lines using submarine and underground cables, fiber optics, satellite connections, etc. Does this mean that the geography of the associated major nodes has been shattered? We shall see that, on the contrary, the cumulative processes and the resulting concentrations have never been so present, even if the technical and organizational specificities of these new networks are leading to a certain remodeling in the geography of key transportation nodes.

1.2. Transportation in the pre-industrial era: organization of long-distance land routes and supremacy of waterways

While the transportation techniques used between the Neolithic and the Industrial Revolution of the 19th century may seem rudimentary, when compared to those we know of today, it is clear that they made significant progress as humanity has never ceased to invent processes to improve speed or load capacity and smooth the roughness of physical spaces. The invention of the wheel at the end of the Neolithic period, the new techniques for surfacing roads (logs, stones or shells from the Neolithic period, paving of certain routes from Roman Antiquity) and the improvement of carriages (making progress in suspension, making materials lighter) made it possible to transport heavier carts, even if the speeds remained very slow: no more than 50 km per day for travelers, about 30 for convoys of goods or livestock, and 5–10 km in the case of steep slopes or up rivers (Bretagnolle 2009). In general, water transportation is preferred to land transportation because it is faster and cheaper, although major trade routes dotted with postal relays have existed since ancient times, such as the Silk and Spice Roads between East and West (Grataloup 2007). In addition to the basic techniques of travel by land (dirt roads, long-distance routes made up of stages spaced out by one day's journey) and water (towing along canals or large rivers, coastal cabotage, major maritime routes dotted with fortified ports), three major innovations in the field of transportation have had a significant impact on the geography of the dominant places in the major regions of the world. These are the postal relay system, long-range maritime transportation (especially transatlantic transportation) and inland navigation via canals and locks.

1.2.1. *Invention of speed by the postal relay system*

The high-speed transportation of information was a privilege reserved for governments in very early periods, as evidenced by the refusal of the French monarchy in 1775 to create a public estafette service (Bretagnolle 2009). Speed was achieved by the frequent change of horses, in relays established at relatively short distances along a route, that is, 10–20 km depending on the time and region. The postal relay system is very old: for example, there were more than 1,500 relays in imperial China in the seventh and eighth centuries before our era. The system set up under the Roman Empire also used relays along strategic routes and benefited from a road network of exceptional quality for the time. Genghis Khan's Mongol Empire (13th century CE) developed a set of relays, which were spaced a day's journey apart, and Northern Italy established its own system of postal roads at the same time. The French monarchy equipped France from the 16th century, but it was mainly in the second half of the 18th century that the network was expanded, with 24,000 km of roads and 1,400 relays in 1783. Postal roads dotted the east coast of the United States in the late 18th century and allowed for the rapid transmission of information (newspapers and letters) between the main cities of the 13 colonies.

Far from being inert, the postal relay system was changing profoundly, not only technically, but also organizationally. In France, for example, speed obtained along these roads doubled in the second half of the 18th century and finally reached about 15 km/h in the 1830s, thanks to the improvement of road paving, lighter carriages, the selection of more efficient horse breeds and the drastic reduction in break times. The choice of routes was also the subject of major revisions. Until the first half of the 18th century, the roads adapted to the relief by preferentially following the main valleys, even if they sometimes crossed mountainous areas, such as the heart of the Massif Central. The main intersections did not necessarily serve the cities, but could be found in the middle of the countryside. All of this changed in the second half of the 18th century, when a co-evolution[1] began to take hold between transportation networks and city networks, and continued to do so. Thus, during the major closures of postal relays brought about by the French Revolution and the wars of the First French Empire, rural crossroads almost permanently disappeared from the landscape. From the 1810s onwards, the most accessible places were big cities, and the roads systematically avoided mountainous terrain (e.g. the Massif Central), because the slopes slowed down the horses. A systematic quest for speed and adaptation to the urban hierarchy began to be characteristic of the choices made by the central administration of La Poste (the postal service) to open relays, in the face of numerous requests from municipalities, which were well aware of the economic issues related to the accessibility of trade routes (Bretagnolle and Franc 2017). Thanks to advances in the speed of the postal carriages, the spacing between the relays was slightly increased (from about 10 km to about 12 km) at the cost of the closure of the former relays, and those located in the smallest localities were removed. The major nodes for the transportation of information and luxury goods by postal roads became the big cities.

1.2.2. *From cabotage to transatlantic navigation (15th–18th centuries)*

Innovations in maritime transportation played a significant role in the first phase of globalization in the 14th and 15th centuries. Fernand Braudel likens this period to a "transport revolution", made possible by a combination of several technical inventions. First of all, that of the strake hull, which gave ships greater strength and the ability to withstand high winds during bad weather, or to sail away from the familiar coasts of the Mediterranean or the Atlantic (Braudel 1979a, 1985). The stern rudder, on the other hand, improved the boat's maneuverability, and the astrolabe enabled it to find its way around in space. Even if the trade flows were

1 The notion of co-evolution refers to the strong interdependencies between the hierarchy of city systems (large, medium-sized and small cities) and the form of transportation networks (levels of accessibility, ie high, medium and low) (Bretagnolle 2009).

extremely small (no more than six or seven connections per year between Lisbon and the main trading posts of the Portuguese Empire), the economic and political repercussions on the cities that chartered the boats were immense. In Western Europe, merchants and entrepreneurs established in the big cities enjoyed relative autonomy from the central governments, insofar as no political or religious authority had managed to impose itself alone in this part of the world: the "city-states" were small worlds where merchants were masters in their own right (Grataloup 2007, p. 114). Emulation and competition between the colonizing metropolises constituted a powerful driving force behind the construction and maintenance of a true world economy[2]. The situation was very different in China, where the fleet was of better quality, but where the political power of the Ming dynasty chose to consolidate land defense at the expense of maintaining the maritime empire that had been established in the first half of the 15th century by Zheng He's explorations.

This first phase of globalization propelled to the top of the urban hierarchy a set of places of election and accumulation, described as world cities by Fernand Braudel (1979b). They held a hegemonic position in trade as they were considered to be the driving force of resource accumulation of that time. The Italian centers of the 13th–16th centuries (Venice, Milan, Genoa and Florence) were gradually being supplanted by Belgian, Dutch and then British competition in the international trade of manufactured goods, in the provision of banking services and in shipping (Pinto 1997). For example, Antwerp, the center of maritime trade after the shift of the world economy from the Mediterranean to Northern Europe, collapsed at the end of the 16th century when the United Provinces decided to close the river Scheldt, during the conflict which pitted them against Spain. Amsterdam then became the hub of international traffic (Hélin 1997), based on the superiority of the Dutch fleet and its role as an import-export warehouse between Mediterranean products and those of the Baltic and the North Sea. From the middle of the 18th century, London acted as a global warehouse, specializing in re-export trade (Chassaigne and Esposito 2013), even before benefiting, from the end of the century, from England's pioneering role in the Industrial Revolution.

Transportation thus plays a key role in this dominant position, and this plays out at several levels. On a global scale, the mastery of maritime transportation constitutes, as we have seen, the first condition for the emergence of a world city. The fleet concentrated in each of these cities was considerable: the scarcity of wood for the construction of ships was one of the main hypotheses put forward by Fernand Braudel to explain the shift of the world economy from northern Italy to the cities of

2 The world economy refers to "a fragment of the universe, a piece of the planet that is economically autonomous, capable for the most part of being self-sufficient and to which its internal connections and exchanges confer a certain organic unity" (Braudel 1979b, p. 14).

the Hanseatic League, in Belgium, then to the Netherlands and England, which were better supplied with timber thanks to the numerous forests of northern and north-eastern Europe. Each of these cities also established a network of trading posts (most of which were fortified and guarded by a military escort), not only in the Mediterranean, but also in the Atlantic and the North Sea, like the Republic of Venice in the 12th century. At the macro-regional level, interurban exchange networks were woven between these metropolises indirectly, thanks to relays established from one area to another and the development of exchange and credit techniques. The fair towns of Champagne or those of the German "Milky Way" were thus valuable meeting places to exchange linen or luxury goods between Southern and Northern Europe. Finally, at the regional level, world cities were surrounded by a cluster of secondary cities with which they maintained complementary relations, for example, in the case of weaving activities. The great water routes of the Brenta, Po and the Adige converged on Venice, along with canals and simple roads (Braudel 1979a, 1985).

1.2.3. *Canals and locks at the nation-state level (16th–19th centuries)*

Until the 16th century, the transport of goods by water was limited in scope, as it depended on topography. Transportation networks were organized within each watershed, with boats adapted to each type of waterway. The construction of towpaths made it possible to slightly improve speeds by facilitating human and animal traction, but slowness remained. Thus, along the Rhône, it took about 1 month to travel the 200 km between Avignon and Lyon. Despite these limited performances, the proximity of navigable waterways allowed some cities to establish their commercial influence, such as Toulouse, Albi or Troyes in France. In other European countries, a system of canals connecting production (e.g. textiles) or consumption sites and the main river or coastal ports propelled certain cities such as Venice, Rotterdam or Amsterdam to the forefront.

At that time, the waterway did not yet constitute a network in its own right: in the case of relief, goods could be moved from one watershed to another only by transiting by land transportation or mule, or by sea (Miquel 1993), and the transloading required to connect different watersheds complicated routes and considerably slowed down journeys. From the 16th century onwards, the invention of pound locks[3] made it possible to overcome the topographical barrier: the creation of junction canals, also known as summit-level canals, led to the appearance of a real network of waterways, where goods could now travel from one watershed to

3 Pound locks with fixed chambers are engineering structures that allow the water level to vary between two gates and thus allow boats to cross the reliefs.

another. This innovation brought about a significant change of scale (Le Sueur 1997), which favored the development of coastal and river towns and changed the way in which inland waterway transportation was organized. In Great Britain, for example, pound locks made it possible in the 18th century to create nearly 2,500 km of canals, linking the various sites of the metallurgical industries to the country's main estuaries. In view of the absence of major topographical obstacles, only a few structures were required to overcome the differences in level. The cost of transporting coal was halved – for example, from Manchester (Marx 1993) – and the routes were much more reliable than those along the coast, subject to the vagaries of the weather.

Junction canals also played a decisive role for some cities in North America. Even though the competition was fierce in inland navigation between the St. Lawrence, the French outlet from the Great Lakes to the Atlantic via Montreal, and the Hudson, an English outlet via New York, the St. Lawrence offered the advantage of journey continuity while the route via the Hudson required portage stages. The situation changed in the 19th century, when the State of New York decided to build a canal connecting Lake Erie and the Hudson River, opening in 1825, and divided travel time by a factor of four and the cost of transportation by 10. The benefits for the city were very significant: even though the port of New York was in competition with the port of Philadelphia, further south, the Erie Canal made New York the new gateway to the products of the West. Traffic increased, and with it also the port's revenues and associated activities. The city's population increased 10-fold in half a century, from 123,000 in 1820 to 1.34 million in 1870 (Bretagnolle et al. 2015).

In northern Europe, similar competition began in the 19th century between Antwerp and Rotterdam, each served by canals connecting hundreds of inland ports. Thus, the development of the Albert Canal facilitated the junction between the North Sea (Antwerp) to the Meuse via Liège, while the Juliana Canal, between Maastricht and the Rhine, connected Rotterdam to the Meuse.

The progressive interconnection of waterways across nation-states led to a movement of "generalized channelization" (Roblin 2003). Faced with the development and modernization of canals, navigable rivers were considered as vulnerable links to the continuity of the transportation network: some passages were difficult, towpaths were not always regular and summer dried up some rivers. Rivers were then channeled to ensure a continuity of circulation. This period was also the period of harmonization of inland waterways: the facilities were designed to guarantee the smooth passage of the barges and tugs whose size was increasingly designed to higher standards. In France, a national plan for the development of canals between 1820 and 1860 was an opportunity to standardize the network by introducing a minimum gauge, the Becquey gauge (named after the director general of the Bureau des Ponts et Chaussées who was at the origin of this plan), which was followed by the Freycinet gauge.

The industrialization of inland navigation and the progressive networking of nation-states thus contributed to the dynamism of the commercial functions of cities that benefited from access to the waterways. While the cities located by the sea had undeniable advantages over others (Beyer 2016), inland cities were now also very accessible when they were at the confluence of two rivers or at a junction with a canal (Kunz et al. 1994).

1.3. The industrial and transportation revolution: scaling up, concentration processes and metropolization

With the revolution in transportation, hazards of all kinds – linked to the climate, insecurity, and wear and tear of mounts – were progressively overcome, enabling fast and regular connections between places. The mechanical speed made possible by the steam engine, used from the 1820s and 1830s to propel locomotives, was immediately four times higher than that achieved up until then and continued to increase: about 40 km/h in the 1840s to around 100 km/h at the end of the 19th century. The effects on geography of the key transportation nodes were immediate, consisting, on the one hand, of a change of scale in the representation and use of spaces, and, on the other hand, in the bypassing of the smallest localities.

1.3.1. *Changes of scale in territorial integration processes*

Increasing speed gradually pushed back the boundaries of the territories that the networks operated in (Offner and Pumain 1996). This process of enlarging the scales – scale-up in the network structure (Garrisson 1990) – went hand in hand with a selection of dominant locations and a gradual disengagement from other localities, whether at the global level, the macro-regional level (and especially at the nation-state level) or the level of cities themselves.

On a planetary scale, steam navigation and national or transnational railways now make the dimensions of the "world" coincide with those of the terrestrial globe (Mackinder 1902): the latter is compared to a "cage" whose cables now go completely around (Brunhes 1909), a closed expanse no longer composed of a marquetry of political borders drawn in broad strokes, but nerve centers weaving their economic networks to enclose vast peripheral regions in their nets (McKenzie 1927). The construction of the Suez Canal (1869) and the Panama Canal (1914) have drastically reduced the distances between remote points of the globe, for example, between London and Singapore (from 16,000 to 10,000 km) or between New York and Los Angeles (21,000 to 8,000 km) (Rodrigue et al. 2013). It was possible to cross the Atlantic, from New York to London, in just 5 days in 1920, compared to 42 at the end of the 18th century (Berger 2003), and ticket prices were divided by 9 between 1850 and 1913 (Grataloup 2007). As early as 1881, the British researcher Francis Galton presented the first isochronous map of the world to the Royal Geographical Society: departing from London, the entire globe was now within the reach of travelers, whether they were adventurers, businessmen or simple tourists. Impact on cities with high accessibility on a planetary scale, global metropolises or simple gateway cities, was huge.

The gateway cities (Bird 1977) are cities located at the interface between different regions of the world. Some large medieval ports already played this role (Burghardt 1971), but new gateway cities appeared in the 19th century, especially thanks to the railroads. For example, Chicago, which had the highest railroad nodality in the second half of the 19th century in the United States (18 converging lines in 1880), connected the urban systems of the East Coast, the West Coast and cities along the Mississippi River. Large cities along the West Coast (San Francisco, Los Angeles) or the Gulf of Mexico (New Orleans) were also major gateway cities allowing trade with Asia, for the former, and Latin America for the latter. Johannesburg in South Africa, founded in 1886 on the site of mining deposits, quickly took the lead in the South African city system thanks to its strong railway nodality allowing it to be connected to the rest of the world via the major ports of Cape Town and Durban. In India, the establishment of the railway network in the second half of the 19th century by the English made it possible to ensure that Delhi, an inland metropolis, became a capital on the scale of the whole country, thanks to the transport of products to the major seaports and to England.

The world's metropolises are themselves gateway cities, but have an additional characteristic: they are located in regions that are centers of the world economy and therefore develop economic and political command functions, in addition to their high accessibility in maritime and/or rail transportation networks. They are also called "capitals of the world" by Paul Vidal de La Blache (1922) or "imperial metropolises" by Halford Mackinder (1902). They are junctions or crossing nodes of land, river or sea routes that constitute interfaces between the local and the global: "The welding of continental currents to maritime currents tends to take place in a limited number of specific points which take on the character of *Emporia*. These enormous cities, the characteristic product of our century, are the organs created by the new trade needs. This is where information is centralized, warehouses are formed, and relationships are established. Something colossal and disproportionate is associated with these creations: the tonnage of ships, the dimensions of docks, the agglomeration of shipyards and factories" (Vidal de La Blache 1922, p. 202). London and New York were the most important world metropolises at that time.

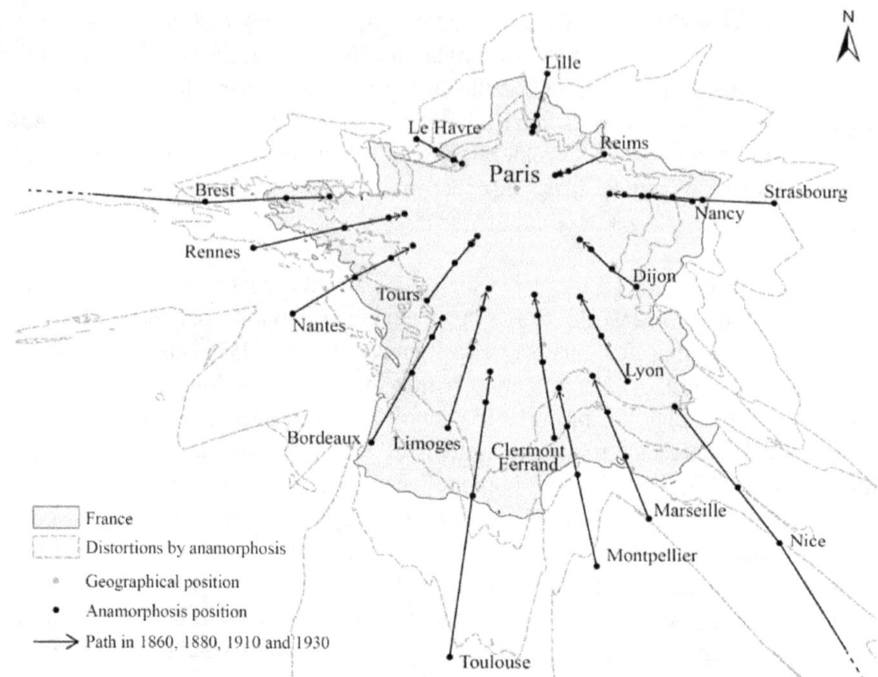

Figure 1.1. *The trajectories of French cities to reach Paris in space-time between 1860 and 1930 (source: Christophe Mimeur (2016)). For a color version of this figure, see www.iste.co.uk/fretigny/transportation.zip*

At the national level, which we speak of here in the case of France, while "the rise of differentiated speed has radically modified the relationship to space-time" (Ollivro 2000), it has nevertheless reinforced the pre-existing hierarchy of places. Between 1860 and 1930, although the French railway experienced uninterrupted growth to reach 50,000 km of track by the latter date (Mimeur 2016), the gains in accessibility, as measured by travel times or network connectivity, were closely correlated with the population levels of the cities served. Paris is the center of the French railway, as programmed by the Legrand law, adopted in 1842, which in fact drew a star network around the capital. Figure 1.1 shows how, between 1860 and 1910, the development of the network brought the main cities closer to Paris. It compares the relative position of cities with respect to Paris, according to topographical (in meters) and topological (in minutes) metrics: the differences between geographical position and anamorphosis position (in space-time) shed light on the role of the railway network in the capacity for exchanges between these cities and the capital. The growth phase of the network describes a permanent acceleration towards Paris, which results in an increasing contraction of space-time. In 1860, the railway network helped to maintain inequalities of access, which were already very visible in the days of the postal network: Brittany and the main mountain ranges are very far from the capital. On the other hand, the development of railway infrastructure from 1880 onwards resulted in a reduction in the gaps between geographical position and position in space-time, which was very clear, for example, for Nice, Brest and Toulouse. From 1910 onward, the network was heavily meshed and changes in accessibility were now marginal. The savings in travel time still benefited the towns that were the most remote, while the development of the network gradually freed itself from the reliefs, although the Massif Central remained aloof, as in the case of the postal routes. The development of the railway network in France in the 19th century clearly reinforced the hierarchy of the city system: Paris is indeed the center of the network, while the main cities were connected at an early stage with speeds that gave them a huge and sudden increase of accessibility.

However, the high rate of growth favored other dominant areas, such as the coastal cities of Le Havre, Bordeaux or Marseille, or cities near the eastern borders, such as Strasbourg or Dijon. Not only did these cities benefit from an access to Paris, which they have retained to this day, but the densification of the network and the optimization of rail speed made it possible to bring these cities closer to each other: in space-time, the distance separating them decreased in relation to their relative geographical position. Indirectly, this structuring of the network also had consequences in terms of accessibility at regional levels. Star networks were organized around the regional capitals, from which radial axes gravitated toward their peripheries (Mimeur and Thévenin 2020). On the other hand, when they existed, the transversal axes were often low-speed infrastructures which did not counterbalance the hierarchy imposed by the axes converging toward Paris. At most, the overall framework of the network co-evolved, in its growth phase, with the urban hierarchy and only local initiatives attempted to correct spatial inequalities by opening transversal lines (Mimeur 2016). However, “as soon as you leave the main lines, there is a marked slowdown” (Studeny 1995, p. 257). Rail innovation was accompanied by major inequalities in service and high-speed differentials. The worst-performing lines, away from the main dominant places inherited from the framework built in the middle of the 19th century, were often the first to disappear after 1930, due to a lack of profitability and having to face road competition that was fast approaching. As a result, 10,000 km of track was removed at the end of the 1930s.

At the city level, the development of rapid urban transportation, followed by highways (from the 1910s to the 1920s in the United States), allowed large metropolises to continue to grow, not only through vertical construction in the form of high-rise buildings (which had been made possible by the invention of the elevator in 1853) but also by the urban sprawl at previously unequalled spans. As early as 1910, the urbanist and sociologist Patrick Geddes described the sprawling growth of London's agglomeration thanks to the railway networks and coined a term (conurbation) to describe this new object, which according to him covered the entire south-east of England.

Sometime later, the automobile, which gave unprecedented individual freedom in the choice of location of residences or businesses, was described by the American sociologist Roderick D. McKenzie (1933) as constituting a genuine transportation revolution. This is associated with the birth of a new geographical object called the metropolitan region, which is now more commonly referred to as a "metropolitan area" or a "functional urban area" (Bretagnolle 2015). The term *metropolitan* suggests a certain dominance of the central municipality on the secondary peripheral centralities, of which there is a trace, for example, in the regional distribution area of newspapers printed in this central city. This is one of the indicators chosen by McKenzie to delineate the Chicago metropolitan area on a map, encompassing some 20 secondary urban centers, the other indicator being the time it takes to travel by train to and from the Chicago station, which when lower than 1.5 hours round trip allows the establishment of commuting (Bretagnolle 2015). From the 1960s in the United States and the 1970s in Europe, the commuter rate was the sole criterion used to delimit metropolitan areas (e.g. in France the Urban Areas in 1996 or the Attraction Areas in 2020).

1.3.2. *Hierarchical selection and bypassing of former stages*

Increasing speed, but also that of the maximum load carried and the progress in the management of schedules, benefitted large cities, but, conversely, led to the bypassing of former, generally smaller nodes, the maintenance of which would have reduced performance, due to the time required for braking and parking (trains) or for stopovers (boats).

The maximum length of wooden-hulled boats was replaced by the "quasi-unlimited" length of metal hulls (Vallaux 1908, p. 276), and ports had to constantly invest in modernizing their equipment and adapting to new technologies. This entailed considerable costs and some of which could not be met (Bretagnolle 2016). Traffic in the 10 largest ports on the north-west coast of Europe grew from 3.9 million tonnes in 1860 to 44.6 million in 1914 to the detriment of the ports of the

South Atlantic, the Mediterranean and the Baltic Sea (Konvitz 1994). On the railroads, the bypassing of cities was linked to the search for greater speed. While the entire urban hierarchy was served by a train station in the 1910s in France, there were notable differences between express lines, which generally ran at more than 80 km/h and served major cities, major tourist resorts and some cross-border towns, and omnibus lines, which ran at around 40 km/h and stopped in small towns. Vallaux (1908, p. 300) observed that "now, thanks to steam, we can skip these ancient stages, and on land, express trains burn the small stations. Any acceleration of transportation, as Ratzel puts it, removes intermediaries that have become useless".

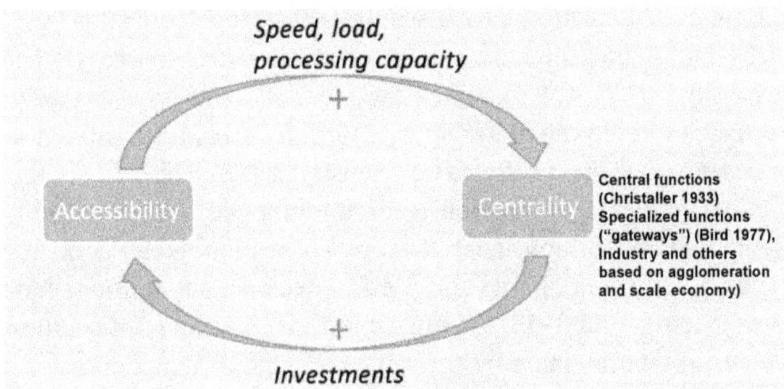

Figure 1.2. *Feedback loop between accessibility and centrality. Adapted from Bretagnolle (2016)*

Figure 1.2 shows the positive feedback loop that was established, from the Industrial Revolution, between the level of accessibility in transportation networks and the level of centrality. Thanks to the increased speed and traffic capacity (goods, passengers), large cities could increase the spatial range of their economic functions. In turn, these gains could be used to further modernize rapid transit infrastructure and improve their accessibility.

Several types of activities are involved. First of all, according to the model developed by Walter Christaller in 1933, the "central" functions refer to the activities of serving the local and surrounding population with goods and services (commerce, administration, etc.). The specialized functions are those of the gateway cities or industrial cities, which are organized into networks (suppliers, customers, subcontractors, subsidiaries, etc.). There are also communication functions (advertising, social media), training and innovation, as well as legal media, whose ranges of attractivity, which are largely virtual, have effects that modify the conditions of material production. Cities that are becoming more accessible thus increase the range of their polarization and their attraction, and ultimately their wealth. In this feedback loop, the ancient and bypassed stages are greatly affected. They are gradually being captured in the areas polarized by the larger ones and are consequently experiencing a decrease in their production, service and stage activities (maintenance, storage, reception of transportation professionals, processing or semi-processing of transported products, etc.). For example, between 1890 and 1925, the total number of ports in the world was reduced by almost 10% and the losses concerned in particular the ports of the Mediterranean and the South Atlantic (Ducruet and Marnot 2015).

The coupling of selection processes from the top (upscaling and territorial integration by metropolises polarizing larger spatial attraction areas) and from the bottom (bypassing of ancient stages on the most efficient transportation networks) contributed to reinforcing the hierarchization of city sizes resulting from the search for economies of scale in the fields of industry and services (agglomeration and urbanization economies)[4]. Population records multiplied: between 1850 and 1913, London went from 2.4 million inhabitants to 7.3 million, New York from 515,000 inhabitants to 4.8 million, Paris and Berlin reached 4.8 million and 4 million inhabitants respectively in 1913. At the same time, while the populations of small towns were generally growing over the same period, they experienced much smaller relative gains than those of large cities, as shown by the concentration indicators of the urban population calculated, for example, from the rank-size distributions of cities – see for the United States, Europe, South Africa or Brazil (Cura et al. 2017). The notion of metropolization, a neologism coined at the end of the 1980s in France to designate the concentration of the most dynamic activities in very large cities, in connection with the acceleration of international openness and current globalization, was visible as early as the second phase of globalization, the end of the 19th century (Bretagnolle et al. 2020).

1.4. Globalization today: a new geography of dominant places shaped by transportation networks?

The current phase of globalization began at the end of the Second World War. Like the previous one, it was characterized by the adoption of public liberalization policies and deregulation and by increasing speeds, reducing transportation costs and increasing load capacity. Three main innovations in transportation and

4 Agglomeration economies refer to the direct effects of spatial clustering in terms of economic efficiency, while urbanization economies are associated with indirect effects: sharing of infrastructure, of the labor market, of services, etc. (Veltz 2008, pp. 21, 73).

communication technology are particularly associated with this current phase of globalization: container shipping, air transportation and the Internet.

As early as the 1960s, many people predicted the emancipation of humanity from the constraints of distance and the roughness of space, reactivating the discourses already present during the transportation revolution (generalized mobility, dispersion of businesses in the countryside, networked world transcending the old scales of the local and global). But quite the opposite, researchers showed the persistence of the old hierarchies (Bretagnolle and Robic 2005). The major nodes associated with the container ships network, airlines network and the Internet remain strikingly identical to those of the second globalization, although some changes are noticeable, associated with new metrics and technical constraints in traffic networks or emerging economic specializations.

1.4.1. *Containerization and international maritime freight: ever-renewed competition between the world's metropolises*

Containerization was invented in the United States in 1956 and enabled goods to be transported in boxes with standardized dimensions. This facilitated international trade. Thanks to this process, maritime transportation became "the backbone of globalization" (Frémont 2007). Today, it forms about 90% of the volume of trade. Load capacity has doubled since the early 1960s and energy consumption reduced by approximately 90% (Bavoux et al. 2005).

The most efficient container ships are now nearly 400 m long, which requires particularly heavy equipment to accommodate these giants of the seas and allow loading or unloading relatively quickly. In 1972, Singapore set up a container terminal (Tanjong Pagar, near the center of the city), but in the 1990s added the Keppel and Brami Island terminals, a few kilometers to the west, before the Pasir Panjang terminals, built between 1998 and 2009 even further west, doubled the island's capacity for container ships. Other major international ports, such as Shanghai, Rotterdam, Antwerp and Hamburg, have also invested astronomical sums to modernize their equipment and build new quays. However, not all ports in the world have the money to carry out this intense work. In half a century, the processes of concentration and hierarchization of port cities have considerably accelerated.

Thus, the race to equip ports with expensive infrastructure has gradually favored the structuring of the maritime network in hub and spokes, with a small number of ports with very high connectivity (working as redistribution platforms) and a large number of edge nodes, served by cabotage or feedering (Varlet and Zembri 2010). This hub-and-spoke model consists of serving as many points as possible from a central point. These points are not directly connected, which sometimes requires a considerable lengthening of distances. Ports specialized in transloading have been developed, for example, Port Kelang (Malaysia) in the Strait of Malacca, Kingston (Jamaica), Gioia Tauro (Italy), Malta and Tangier Med (Morocco).

With the globalization of the production system and the concentration of the manufacturing industry in East Asia, the Atlantic ports (Europe, West Coast of the United States) have largely fallen behind the Pacific ports (Table 1.1).

Rank	Port name	Country	Traffic (thousands of TEUs)
1	Shanghai	China	40,233
2	Singapore	Singapore	33,666
3	Shenzhen	China	25,208
4	Ningbo-Zhoushan	China	24,607
5	Hong Kong	China	20,770
6	Busan	South Korea	20,493
7	Guangzhou	China	20,370
8	Qingdao	China	18,262
9	Dubai	United Arab Emirates	15,368
10	Tianjin	China	15,040
11	Rotterdam	Netherlands	13,734
12	Port Kelang	Malaysia	11,978
13	Antwerp	Belgium	10,450
14	Xiamen	China	10,380
15	Kaohsiung	Taiwan	10,271
16	Dalian	China	9,707
17	Los Angeles	United States	9,343
18	Hamburg	Germany	8,860
19	Tanjung Pelepas	Malaysia	8,260
20	Laem Chabang	Thailand	7,670

Table 1.1. *The 20 largest ports in containerized traffic in the world in 2017 (sources: International Yearbook (CIY) 2018, data presented on the IAPH [International Association of Ports and Harbors] website). TEU: twenty-foot equivalent (equivalent to the volume of a standard container)*

Although accessibility indicators related to maritime transportation are never mobilized by the researchers establishing the world ranking of the world's largest metropolises, described since the work of Peter Hall, John Friedmann or Saskia Sassen as world cities or global cities (Bretagnolle et al. 2020), the maritime issues still remain extremely high. World cities are home to an impressive number of residents and businesses. Accordingly, the upholding of an autonomy in the routing of physical flows and securing access to the resources necessary for their existence have motivated the large-scale investments not only in the ports of Asia, but also in those of the American continent, Europe, Africa and the Middle East.

1.4.2. *Air: a specific geography of dominant spaces and places*

Air transportation is the most recent innovative industry, since it was not until 1914 that Tony Jannus covered the 27 km commercial flight from St. Petersburg to Tampa, Florida, in 23 min. One hundred years later, the growth of air transportation is still exponential: traffic has doubled every 15 years to reach more than 4 billion passengers per year in 2017, despite numerous crises such as the oil crisis in the 1970s, the attacks of September 11, 2001 or the financial crisis of 2008, which are certainly visible in curves, but with a limited impact over time.

Although the first airlines were first developed within the national States, technical progress quickly made it possible to establish international connections, particularly between the colonies and metropolitan areas. The first international regulations were adopted at the end of the Second World War, with the Chicago Convention (1944), followed by a series of bilateral agreements establishing not only freedom of air travel, but also technical, economic and diplomatic regulations. In theory, air transportation promotes "generalized exchange between the different parts of the planet, the world space being then the transaction space of humanity" (Dollfus 1997). In reality, flows, both of goods and passengers, reflect a very strong polarization, with a concentration of routes between Europe, North America and Asia-Pacific. Thanks to the lighter weight of materials and advances in propulsion techniques, travel requires fewer and fewer intermediate stages, thus strengthening some airport cities at the expense of others. But deregulation, which has been operating since the 1970s in the United States and more recently in Europe (Varlet 1997), is gradually establishing a dominant air network model, organized in hub and spokes (i.e. star-shaped, see above) and no longer point-to-point. It should be noted, however, that this model is applied to different geographical configurations depending on the focus adopted – "there is no standardized statistical definition of a hub" (Le Goix 2016) – which means that the hierarchy of dominant places varies according to the indicators used.

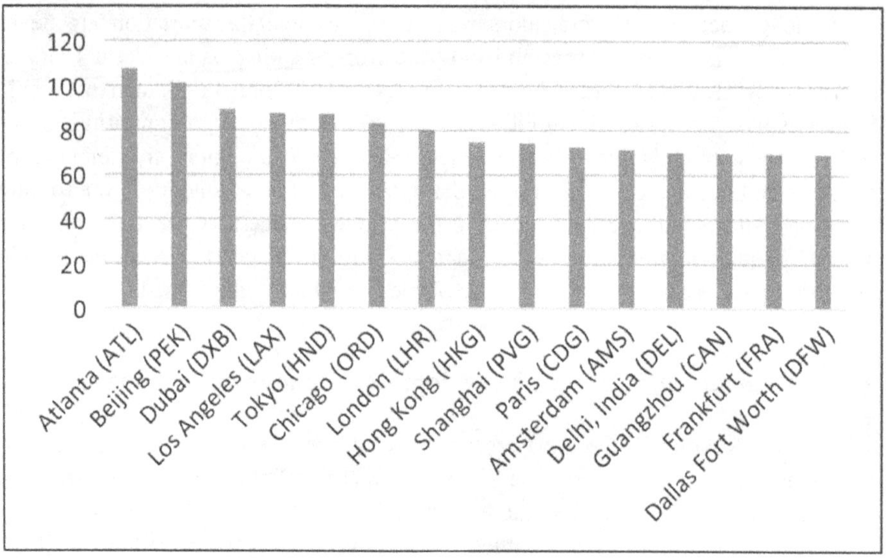

Figure 1.3. *Number of passengers in various major airports in 2018 (in millions) (source: Airports Council International)*

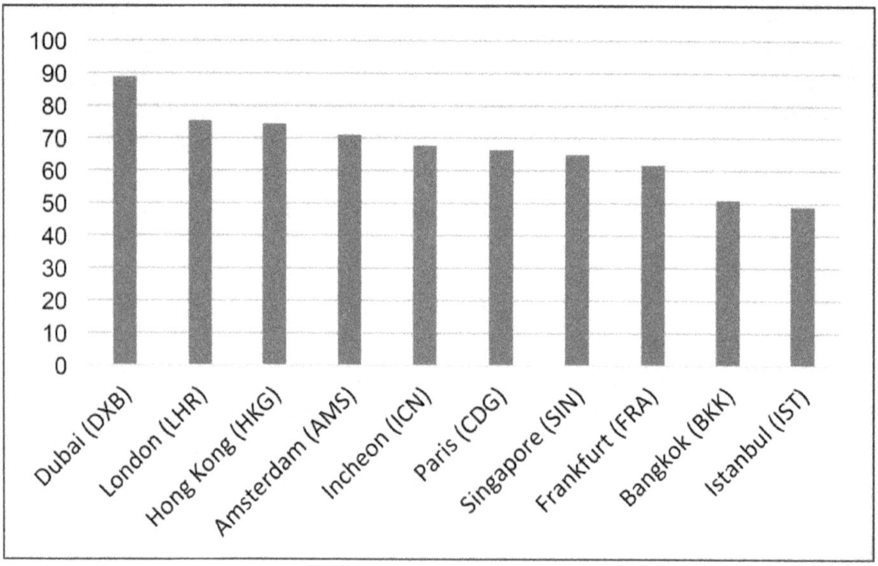

Figure 1.4. *Number of international passengers in various major airports in 2018 (in millions) (source: Airports Council International)*

If we look at the number of passengers, there are strong differences between international passengers, transit passengers or passengers on domestic or intracontinental flights. Indeed, air alliances[5] have made it possible to gain market share by maximizing connections: each scheduled airline has its own hub, from which it can operate its own routes and draw passengers from partner airlines. The global hierarchy of airport metropolises mainly reflects this hub and spokes model of organization. Thus, while Asian metropolises are extremely dynamic with Beijing, Tokyo, Hong Kong and Shanghai, as for maritime transportation, New York City is not among the top cities in the United States (Figure 1.3). It is Atlanta (Georgia), only the 9th largest city in terms of total population, that received the most passengers in 2018 and is even the world leader, not only because it receives a huge number of international passengers but also because it serves as a connecting hub for Delta Air Lines (Figure 1.4). The presence of this Delta Air Lines hub would have allowed the city to assert its status as a metropolis, not only regionally but also globally (Whitelegg 2000). The geography of passenger airports therefore combines dominant positions in the international market and strategic cities for the dynamics of domestic markets (Matsumoto 2018). While the historical structuring of national airlines partly explains the prominent positions occupied by European cities such as Paris, London and Frankfurt in the global hierarchy of airports, the dominant locations in the American domestic market are, thus, the result of the internal organization of the airlines: Atlanta is the Delta hub, Chicago the United one and Dallas the American Airlines one. More surprisingly, Dubai climbed to third place in 2018, while its airport only entered the top 20 rankings in 2007. Emirates' strategic positioning as a hub between Europe and Asia and between Africa and Asia makes Dubai a highly visited airport, which is mainly due to transit passengers.

If we look at goods this time, flows are dominant between the world's three largest hubs: Europe, North America and Asia-Pacific. These flows thus follow the general trend in international maritime trade and the organization of hub and spokes is particularly effective (Frémont 2016). The major Asian airports of Hong Kong and Shanghai (Figure 1.5) overtook US airports in terms of cargo volume between 2014 and 2018. But the geography of airports is not necessarily the result of leadership functions associated with large metropolises, as is the case for passengers. Thus, the largest air freight hubs in the United States are taking

5 Most of the national airlines have been grouped together in alliances since the end of the 1980s. Skyteam, Star Alliance and Oneworld are the three main alliances that practice "code-sharing": within the same alliance, airlines can sell route sequences operated by their partners. By maximizing possible destinations, even if it sometimes requires the use of a partner airline, incumbent airlines also see it as a way to build customer loyalty.

advantage of their localization: FedEx and UPS are established in Memphis and Louisville, respectively. Centrally located in the country, these hubs have been modernized thanks to computer tracking and automation of logistics operations to redistribute packages and letters as efficiently as possible. This location also makes it possible to free themselves from the traffic-related constraints of passenger aircrafts. However, in Europe, these same large companies tend to be located in an airport close to a metropolis, probably because space constraints are greater than in the United States.

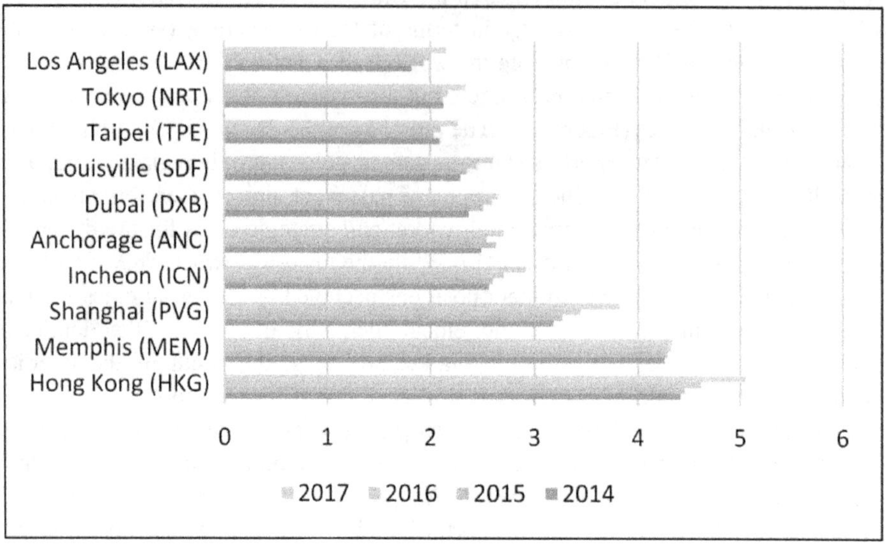

Figure 1.5. *Cargo volume airports around the world from 2014 to 2018 (million tonnes) (source: Airports Council International). For a color version of this figure, see www.iste.co.uk/fretigny/transportation.zip*

Thus, while the geography of the dominant places in passenger air transportation is characterized by a concentration of flows in an increasingly small number of airport cities, the aerial cargo approach offers a nuance to these repeated iterations. However, the gradual increase in deregulation of air transportation has resulted in the emergence and strengthening of secondary cities: "Liberalization of the skies benefits the cities and higher-ranking regions, as the main transmitters and receivers of air flows and through the strengthening or development of hubs; however, it also benefits certain mid-or lower-tier cities or regions that manage to succeed through alternative strategies" (Dobruszkes 2008, p. 35). Low-cost traffic development, while it does not shake up the global hierarchy, proposes alternatives to hub and spokes: it is possible to reach point A and point B without transiting through a huge airport serving as a hub to a national airline, thus representing an alternative point-to-point model. The main objective of these airlines is to rationalize their operating costs by establishing bases from which connections are limited to optimize rotations. For example, this model is well established in Europe, although its geography has evolved since the 1990s (Dobruszkes 2006). The first destinations were located inland of the European continent, then increasingly toward the Mediterranean and Adriatic coasts. Beginning in the mid-2000s, many low-cost European airline companies opened routes to North Africa and Turkey, confirming the fact that the tourist clientele was their preferred target. The situation was different in the United States: although deregulation led to the appearance of new entrants in the market at the end of the 1970s, it very quickly implied a concentration of stakeholders, who most often divided the market by major regions. The point-to-point model was not the rule and very quickly it slipped into a new model based on hub and spokes (Zembri 2016): for instance, the company Southwest created new dominant places and moved closer to the traditional strategies of regular airlines.

It should be noted that air transportation is one of the fast-moving sectors, with infrastructures limited to network access points. It is only the capacity of these entry and exit points that can limit the freedom of companies, but competition remains very strong. This is evidenced by the rapid growth of airports in the Middle East, as well as the total reorganization of the European airlines offers when low-cost airlines appeared in the 1980s.

1.4.3. *The material geography of the Internet: data transmission and storage infrastructures*

As early as 1933, the American sociologist Roderick D. McKenzie described the world as a gigantic spider's web woven thanks to advances in telecommunications, which now made it possible to transmit intelligence directly, whereas information had until then depended on transportation networks, whether road, rail or maritime. However, it was not until the 1960s that a real revolution in information delivery technologies took place. The telegraph and the telephone were certainly important, but they did not allow for the transmission of voluminous information, especially written documents, which were the basis of the management and administration procedures of companies. The revolution in transmissions and telecommunications marked a turning point, highlighted in 1964 by Melvin Webber, a professor of urban planning at the University of California, Berkeley. In addition to transmitting information "infinitely" and in "unlimited" quantities (properties already emphasized with regard to the telegraph a century earlier), "cybernation" made it possible to ensure the "automation" of information and the "control" of informative remote tasks" (Webber 1964). The impacts of this new type of infrastructure network on major nodes of transportation and communications were analyzed as early as the 1960s by several researchers. In 1966, the Swedish geographer Torsten Hägerstrand emphasized the importance of studying the "information fields" through which ideas circulate and their diffusion in the urban hierarchy, taking the example of the expansion of Rotary clubs in Europe in the 1920s and 1930s (Hägerstrand 1966). A few years later, Pred (1973, p. 10) affirmed the primary role of information flows in the organization of city systems: "In recent years, many authors have commented both on the declining role of transportation as a determinant of localization and on the increasing impact of information". Due to the massive internationalization of exchanges and the structuring role of information flows, which are intangible and therefore particularly difficult to grasp, research on contemporary city systems is now approached as a particular field of complexity sciences. In contrast to the city systems that were largely based on the perimeters of nation-states in the 19th century and until the 1970s, it is now difficult, if not impossible, to define the perimeters of current city systems and the complex co-evolutions that take place with transportation and communication networks.

Nevertheless, we can draw a few observations based on the materiality of the infrastructures that are necessary for the operation of the Internet network. First of all, the vast majority of these infrastructures consist of undersea cables. Thus, 90% of international data uses these lines, despite the rise of satellite communications from the 1970s onwards. The arrival of fiber optics in the 1980s and 1990s reactivated interest in submarine cables, which are extremely concentrated in space because they adapt to the geography of the world's metropolises and the centers of the world economy. Three major routes stand out: the transatlantic axis, the transpacific axis and the Europe-Asia axis, which passes through the Suez Canal. The map of international routes of the global internet network in 2018 (Figure 1.6) also reveals the presence of areas of strong interconnection on a macro-regional scale, for example, at the scale of Asia or Europe. On the other hand, international connections within Africa and Latin America are weak compared to those to the centers of the world economy, reflecting the growing disjunction between places of consumption and standardized information processing facilities. The latter are engaging in multiple activities: phone bookings, digitally provided commercial services, call centers and hotlines, after-sales services relocated abroad, for example, to the Maghreb for French-speaking countries or to India for the English-speaking world (Bretagnolle et al. 2011). Another important aspect of the materiality of the infrastructures necessary for the functioning of the Internet network is the presence of large digital data storage warehouses, called data centers, which enable the provision of services from cloud computing. Like undersea cables, they are highly concentrated in the world and reflect the geography of the world's metropolises and centers of the world economy with a particularly strong presence in North America, Western Europe and, somewhat more secondarily, in East Asia (Figure 1.7). At the sub-metropolitan level, data centers contribute to urban sprawl because of their preferential location on the periphery, for example, in the case of Paris city-region, on the Saclay plateau and Plaine Saint-Denis, and, in the United States, in Silicon Valley, south of San Francisco (Diguet and Lopez 2019).

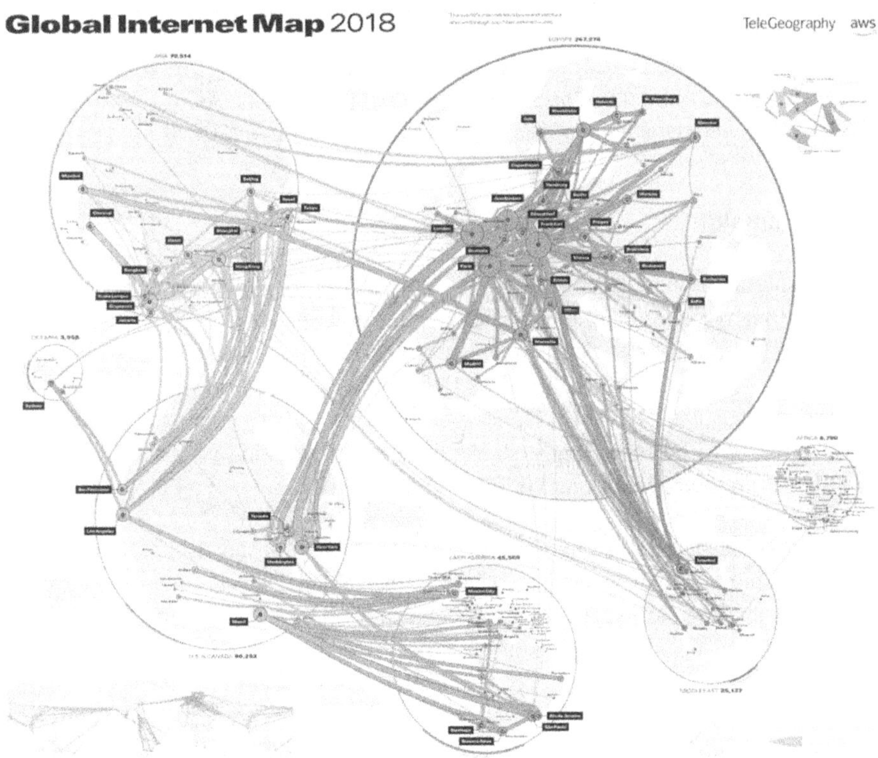

Figure 1.6. *International Internet roads in 2018 (source: telegeography.com). For a color version of this figure, see www.iste.co.uk/fretigny/transportation.zip*

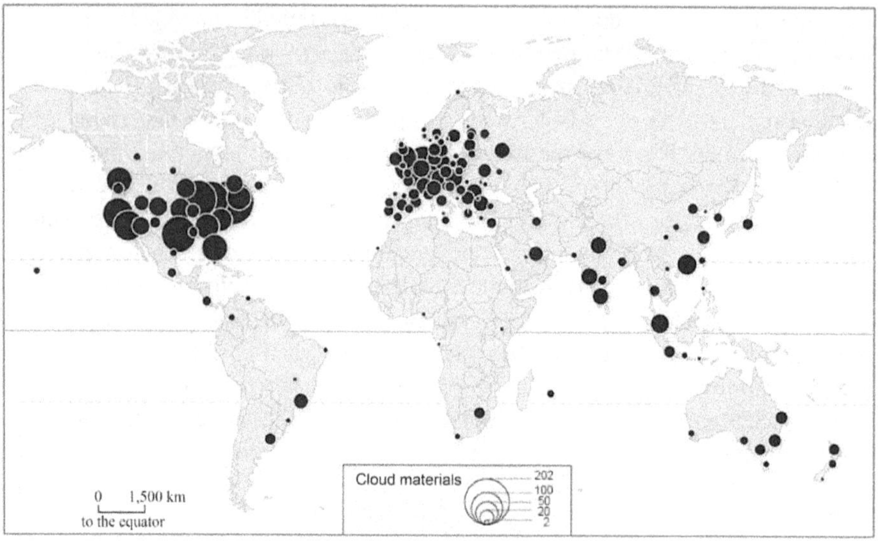

Figure 1.7. *The distribution of data centers as cloud computing materials (source: Laurent Carroué (2019))*

1.5. Conclusion

Transportation infrastructure networks are enduringly embedded in space to the point that many researchers have tried to identify a spatial development model of their own (Offner 1993). The superposition of networks over the long term thus shows their "permanence" (Dupuy 1991): when a new network appears, old ones rarely disappear entirely. This chapter also shows that the geography of dominant places appears to be surprisingly stable from the networking of the world's great cities by sea to their networking via fiber optics.

The dominance of these major nodes can be seen at all levels of scale: what is true at the international level is also true at the national level, where networks of waterways and then railways and roads contribute to the selection of certain places to the detriment of others, which are bypassed. At the scale of the city, this same network accumulation first affects the center before spreading unevenly to the periphery. However, organizational innovations related to certain types of networks explain exceptions to the permanence of key transportation nodes: transshipment in the case of maritime transportation or the logic of the hub in the case of air transportation.

Over centuries, network-related innovations have repeatedly led to a "speed revolution" (Studeny 1995), resulting in major changes in the relationship with time and space. Thus, while networks have allowed a significant increase of interactions between places, the positions of these places in space-time, on the one hand, and in geographical space, on the other hand, have sometimes been completely dissociated (Bretagnolle and Robic 2005). In addition, the opening up of the world's spatiotemporal horizon has been accompanied by a strong hierarchy of places to the point of distinguishing premium spaces (Graham and Marvin 2001), which have access to high-performance and expensive networks, while other areas still have little or much less efficient infrastructure. Faced with the homogenization of the world announced by some Saint-Simonians in the 19th century, research shows, on the contrary, that the growth of networks has reinforced pre-existing inequalities or even created new ones (Dupuy 2011). The coexistence of the different networks is also an opportunity for competition or complementarity, inducing, for some, logics of self-segregation or club effects and, for others, space relegation.

The geography of dominant transportation places thus reveals critical access points and lines both internationally and locally. Major airports, major trade routes and large data warehouses are all strategic axes and points in the organization of the world today. Geopolitical considerations have always accompanied them and have even been reinforced as the network's concentration has intensified. This is evidenced by the Suez Canal crisis in 1956 or, more recently, the issues related to the New Silk Road between Asia and Europe (Astarita and Damiani 2016; Thorez 2016).

While the question of the nature of the interactions between transportation and territory continues to be raised by the scientific community, the geohistorical perspective fleshes out the main features of a close co-evolution between the dynamics of transportation networks and those of urban networks and city systems in the world. In this co-evolution, major transportation places are impacted, as well as secondary or bypassed places.

1.6. References

Astarita, D. (2016). Géopolitique de la nouvelle route de la soie. *Géoéconomie*, 79(2), 57–94.

Bavoux, J.-J., Beaucire, F., Chapelon, L., Zembri, P. (2005). *Géographie des transports*. Armand Colin, Paris.

Berger, S. (2003). *Notre première mondialisation. Leçons d'un échec oublié*. Le Seuil, Paris.

Beyer, A. (2016). Les grands jalons de l'histoire des voies navigables françaises. *Pour mémoire*, 17, 83–93.

Bird, J. (1977). *Centrality and Cities*. Routledge and Kegan Paul, London.

Braudel, F. (1979a [1949]). *La Méditerranée et le monde méditerranéen à l'époque de Philippe II*. Armand Colin, Paris.

Braudel, F. (1979b). *Civilisation matérielle, économie et capitalisme, XVe-XVIIIe siècles. Tome 3 : Le temps du monde*. Armand Colin, Paris.

Braudel, F. (ed.) (1985). La mer. In *La Méditerranée. L'espace et l'histoire*. Flammarion, Paris.

Bretagnolle, A. (2009). Villes et réseaux de transport : des interactions dans la longue durée. HDR Thesis, Université Paris 1 Panthéon-Sorbonne, Paris.

Bretagnolle, A. (2014). Les effets structurants des transports, une question d'échelles ? *L'Espace Géographique*, 43(1), 63–65.

Bretagnolle, A. (2016). City-systems and maritime transport in the long-term. In *Maritime Networks: Spatial Structures and Time Dynamics*, Ducruet, C. (ed.). Routledge, London.

Bretagnolle, A. and Franc, A. (2017). Emergence of an integrated city-system in France (XVIIth–XIXth centuries): Evidence from toolset in graph theory. *Historical Methods: A Journal of Quantitative and Interdisciplinary History*, 50(1), 49–65.

Bretagnolle, A. and Robic, M.C. (2005). Révolution des technologies de communication et représentation du monde : 1. Monde-point et monde difforme (1830–1840). *L'information géographique*, 69, 150–167.

Bretagnolle, A., Le Goix, R., Vacchiani-Marcuzzo, C. (2011). *Métropoles et mondialisation*. La Documentation française, Paris.

Bretagnolle, A., Delisle, F., Mathian, H., Vatin, G. (2015). Urbanization of the United States over two centuries: An approach based on a long-term database (1790-2010). *International Journal of Geographical Information Science*, 29(5), 850–867.

Bretagnolle, A., Rivière, D., Douay, N. (2020). Métropoles, métropolisation et gouvernance métropolitaine. In *Le temps long du peuplement. Concepts et mots-clés*, Sanders, L., Bretagnolle, A., Brun, P., Ozouf-Marignier, M.-V., Verdier, N. (eds). Presses Universitaires de Tours, Tours.

Brunhes, J. (1909). Les limites de notre cage. *Le Correspondant*, 10 December, 833–862.

Burghardt, A.F. (1971). A hypothesis about gateway cities. *Annals of the Association of American Geographers*, 61(2), 269–285.

Carroué, L. (2019). *Géographie de la mondialisation*. Armand Colin, Paris.

Chassaigne, P. and Esposito, M.-C. (2013). *Londres, la ville-monde*. Vendémiaire, Paris.

Christaller, W. (1966). *Central Places in Southern Germany*. Prentice Hall, Upper Saddle River.

Cura, R., Cottineau, C., Swerts, E., Ignazzi, C., Bretagnolle, A., Vacchiani-Marcuzzo, C., Pumain, D. (2017). The old and the new. Qualifying city systems in the world with classical models and new data. *Geographical Analysis*, 49(4), 363–386.

Diguet, C. and Lopez, F. (2019). L'impact spatial et énergétique des data centers sur les territoires. Report, ADEME, Project ENERNUM [Online]. Available at: https://librairie.ademe.fr/cadic/910/rapport_enernum_final-_20_fevrier_2019.pdf [Accessed 29 September 2022].

Dobruszkes, F. (2006). An analysis of European low-cost airlines and their networks. *Journal of Transport Geography*, 14, 249–264.

Dobruszkes, F. (2008). *Libéralisation et desserte des territoires : le cas du transport aérien européen*. P.I.E. Peter Lang, Brussels.

Dollfus, O. (1997). *La Mondialisation*. Presses de Science Po, Paris.

Ducruet, C. and Marnot, B. (2015). Analyser les trafics portuaires mondiaux en 1890 et 1925 à partir des registres du Lloyd's. In *La maritimisation du monde de la préhistoire à nos jours*, GIS d'histoire maritime (ed.). Presses Universitaires de Paris-Sorbonne, Paris.

Dupuy, G. (1991). *L'urbanisme des réseaux : théories et méthodes*. Armand Colin, Paris.

Frémont, A. (2007). *Le monde en boîte : conteneurisation et mondialisation*. Lavoisier, Paris.

Frémont, A. (2016). Transport aérien de fret et mondialisation. *Questions internationales*, 78, 92–97.

Garrisson, W.L. (1990). Networks: Reminiscence and lessons. *Flux*, 1, 5–12.

Geddes, P. (1915). *Cities in Evolution*. William & Norgate, London.

Graham, S. and Marvin, S. (2001). *Splintering Urbanism. Networked Infrastructures, Technological Mobilities and the Urban Condition*. Routledge, London.

Grataloup, C. (2007). *Géohistoire de la mondialisation. Le temps long du monde*. Armand Colin, Paris.

Hägerstrand, T. (1966). Aspects of the spatial structure of social communication and the diffusion of information. *Papers of the Regional Science Association*, 16, 27–42.

Hélin, E. (1997). Les Pays-Bas. Cohabiter dans le même microcosme : Pays-Bas méridionaux. In *Histoire des populations de l'Europe. I. Des origines aux prémices de la révolution démographique*, Bardet, J.-P. and Dupaquier, J. (eds). Fayard, Paris.

Konvitz, J.W. (1994). The crisis of Atlantic port cities, 1880 to 1920. *Comparative Studies in Society and History*, 36(2), 293–318.

Kunz, A., Merger, M., Weil, C. (1994). Navigation intérieure, industrialisation et développement économique dans l'Europe du XIXe siècle. *Histoire, économie et société*, 13(4), 697–701.

Le Goix, R. (2016). Aéroports et métropoles : l'important, c'est d'arriver là où on veut aller. *Questions internationales*, 78, 62–79.

Le Sueur, B. (1997). La voie d'eau, une machine hydraulique, outil polyvalent d'aménagement du territoire. *Annales de Géographie*, 593(594), 195–204.

Mackinder, H.J. (1902). *Britain and the British Seas*. William Heinemann, London.

Marx, R. (1993). *Histoire de l'Angleterre*. Fayard, Paris.

Matsumoto, H. (2018). International air network structures and air traffic density of world cities. *Transport Research Part E*, 43, 269–282.

McKenzie, R.D. (1927). The concept of dominance and world-organization. *The Americain Journal of Sociology*, 33(1), 28–42.

McKenzie, R.D. (1933). *The Metropolitan Community*. Russell and Russell, New York.

Mimeur, C. (2016). Les traces de la vitesse entre réseau et territoire, approche géohistorique de la croissance du réseau ferroviaire français. PhD Thesis, Université de Bourgogne Franche-Comté, Dijon.

Mimeur, C. and Thévenin, T. (2020). Analyse diachronique de la croissance du réseau ferroviaire français entre 1860 et 1930 : entre expansion connexionniste et sélection hiérarchique ? *Flux*, 122(4), 69–87.

Miquel, P. (1993). *Histoire des canaux, fleuves et rivières de France*. Hachette, Paris.

Offner, J.-M. (1993). Les effets structurants du transport : mythe politique, mystification scientifique. *L'Espace géographique*, 22(3), 233–242.

Offner, J.-M. and Pumain, D. (eds) (1996). *Réseaux et territoires. Significations croisées*. Editions de l'Aube, La Tour-d'Aigues.

Ollivro, J. (2000). *L'homme à toutes vitesses : de la lenteur homogène à la rapidité différenciée*. Presses Universitaires de Rennes, Rennes.

Pinto, G. (1997). L'Italie. Le Moyen Âge. In *Histoire des populations de l'Europe. I. Des origines aux prémices de la révolution démographique*, Bardet, J.-P. and Dupaquier, J. (eds). Fayard, Paris.

Pred, A. (1973). The growth and development of systems of cities in advanced economies. *Lund Studies in Geography B*, 38, 9–82.

Roblin, L. (2003). *Cinq siècles de transport fluvial en France : du XVIIe au XXIe siècle*. Editions Sud-Ouest, Bordeaux.

Rodrigue, J.-P. (2013). *The Geography of Transport Systems*. Routledge, New York.

Studeny, C. (1995). *L'invention de la vitesse*. Gallimard, Paris.

Thorez, J. (2016). La nouvelle route de la soie : une notion porteuse d'illusion. *Questions internationales*, 82, 33–41.

Vallaux, C. (1908). *La mer : géographie sociale. Populations maritimes, migrations, pêche, commerce, domination de la mer*. Octave Doin Editeur, Paris.

Varlet, J. (1997). La déréglementation du transport aérien et ses conséquences sur les réseaux et sur les aéroports. *Annales de géographie*, 593(594), 205–217.

Varlet, J. and Zembri, P. (2010). *Atlas des transports*. Editions Autrement, Paris.

Vidal de la Blache, P. (1922). *Principes de géographie humaine*. Amand Colin, Paris.

Webber, M. (1964). The urban place and the nonplace urban realm. In *Explorations into Urban Structure*, Webber, M.M. (ed.). University of Pennsylvania Press, Philadelphia.

Whitelegg, D. (2000). Keeping their eyes on the skies: Jet aviation, Delta Air Lines and the growth of Atlanta. *The Journal of Transport History*, 21(1), 73–91.

Zembri, P. (2016). Le transport aérien low-cost. *Questions internationales*, 78, 59–61.

2

Gateways or Globalization Taking Place(s): Interdependencies in Question

Nadine CATTAN[1] and Jean-Baptiste FRÉTIGNY[2]

[1]Géographie-cités, CNRS, Paris and Aubervilliers, France

[2]PLACES, CY Cergy Paris Université, Cergy-Pontoise, France

2.1. Introduction: staples of globalization

Gateways are pivotal vectors of globalization, offering a concrete grasp of how this puzzling planetary process shapes places and territories. Gateways often refer to very large infrastructures engaged in intense international relations. In 2019, just before the Covid-19 pandemic, airports welcomed 4.5 billion passengers, with 41% of them being international travelers. This overall passenger volume was twice that of 2006 (ICAO 2021). Similarly, 811 million 20-foot equivalent containers passed through seaports in 2019, more than double the figure for 2005 (UNCTAD 2021). These infrastructures reinforce dominant transportation modes and polarizing logics of exchange, deepening the divide between core and peripheral areas – deeply embedded in our representations of spatial hierarchies. This approach to gateways (see van Klink and van den Berg 1998, for example) enshrines a vision that elevates certain places and territories as exceptional, confining our understanding into binary and hierarchical frameworks that overlook the complexity of the processes, relationships and functionalities involved.

As a multifaceted process that transcends any single dimension, globalization demands close attention to the diversity of mobilities – one that captures their full spectrum and embraces a comprehensive, cross-cutting approach to flows. Accordingly, the aim of this chapter is to clarify existing conceptions of gateways and lay the foundations for a renewed theorization. This reflection draws on three sources of material. The first is a critical review of French- and English-language academic literature from the 2000s onward in the fields of mobility, migration and transportation studies. The second is the result of a collective prospective work that explored France's gateways in 2040[1]. The third is based on our recent studies on migrations related to globalization from below as well as on varied large-scale, extra-local mobilities (Cattan 2012; Frétigny 2016; Cattan and Frétigny 2021). At the intersection of these materials, this chapter develops a categorization of gateways in their various forms to enhance understanding of the broad range of exchanges, activities and actors they entail. This typological framework breaks away from the dominant approaches in the literature, which remain largely segmented, either by transportation mode or by travel motive (tourism, migration, work, etc.). Additionally, it aims to incorporate the often-neglected informal dimension of these circulations, at stake in all practices at a distance from the most visible rules and standards, whether or not they cross into illegality.

A gateway is frequently conceptualized as an interface between a national territory and the wider world. This "entry point" – possibly ethnocentric – approach considers gateways as oriented toward a given territory, with flows directed inward. Yet this view does not fully capture the web of interactions that unfold across multiple scales, embedding these gateways in dense networks of relationships between places and between actors. If we ask ourselves which is France's primary port, for instance, we may well need to look at Antwerp rather than Le Havre, highlighting the intense interdependence of gateways and network dynamics, disrupting taken-for-granted national logics.

In this chapter, we consider gateways insofar as they involve international relations and draw on the theoretical reflection encapsulated by the term in the English-speaking literature. The notion of "gateway city" appeared early in geographic and urban studies, with its first reflexive use most likely dating back to the 1930s (McKenzie 1933). Ullman (1957) approached them by focusing on the concentration of the headquarters of most United States railway companies or

1 As part of the Territories 2040 program (Cattan and Frétigny 2011). This program was developed by the national authority in charge of planning (DATAR). Gateways were more precisely conceived by the authority as entry points of the territories ("*portes d'entrée*" in French).

steamship companies. It was only with Burghardt (1971) that gateway cities were distinguished from central places. The central place is associated with population and resource accumulation, as well as the polarization of exchanges, particularly with its hinterland, following an areal logic of proximity. From the 1970s, however, the gateway city has been seen as a point of contact, with its dynamics fueled by connections with distant places and taking advantage of situational differentials, such as a cross-border city, leveraging advantages from each side of the border. While they follow different logics, the gateway city and the central place are not mutually exclusive (Bird 1983, p. 200). In transportation studies, the concept of gateway city has been revisited to analyze the relationship between gateways and corridors (Rodrigue 2004; Prentice 2006): gateway cities are seen as network hubs supervising the operations related to corridors, such as New York for the Boston-Washington corridor or Hong Kong for the Pearl River Delta one. Pain (2011) and Short et al. (2000) advocate for a relational approach to understanding gateways, encouraging us to view them within the context of rescaled proximity relations that are increasingly globalized. This shift has led to the emergence of true "hinterworlds", expansive regions of influence that extend beyond traditional local hinterlands.

How does globalization take place through gateways? This chapter stresses the importance of all actors shaping gateways and of their spatialities, whether when practicing them or when holding responsibility over them, as gatekeepers. Section 2.2 examines the various modalities through which these places are assembled, identifying the components, networks and practices involved. Building on these observations, section 2.3 unpacks the challenges raised by these gateways and their protagonists, from attracting flows to being vulnerable to multiple risks.

2.2. A kaleidoscope of gateways

By considering gateways from a relational perspective across multiple scales, four major types of gateway emerge. The presence or absence of transportation and logistics infrastructures, the diversity of mobilities, the range of activities and the characteristics of the practices involved are four criteria that enable us to grasp what takes place in these globalized gateways. In their various forms, these gateways are always context specific. But they do not merely act as a threshold between their hinterland and a set of privileged international linkages. They also engage a plurality of interconnections that place them in a network of interdependent gateways worldwide.

2.2.1. *Site gateways*

The first category, which we refer to as site gateways, corresponds to the most easily identified gateways, strategically located within transportation networks. They align most closely with the dominant representations of gateways and stand out from other types of gateways. They are distinguished by the strong presence of transportation and logistics infrastructures, such as main railway stations, bus terminals or (air)ports, marked by various mobilities of people or goods. They can also be logistics places, as well as gas or oil terminals, which involve more specialized flows and services.

Beyond the prominent site gateways such as Dubai International and New York J.F.K. airports or Shenzhen and Rotterdam ports, less visible gateways play crucial roles for more isolated regions. Kangiqtiniq airport is a perfect illustration of the significance of these gateways (Figure 2.1) for the Nunavut territory in Canada's Far North. It also serves a major role in conveying tourists from Winnipeg, Manitoba, to Hudson Bay attractions.

Figure 2.1. *Kangiqtiniq Airport (otherwise known as Rankin Inlet), Nunavut Territory, Canada (source: Cattan (2018)). For a color version of this figure, see www.iste.co.uk/fretigny/transportation.zip*

Alongside these technologically sophisticated gateways closely tied to transportation, road site gateways can arise solely from border crossings. Notable examples include crossing points between the United States and Mexico, or between Israel and Palestinian territories, often literal gates within actual walls. In the Schengen zone, the materialization of these road site gateways, which had partially eroded, can be strongly reactivated, as in the context of the Covid-19 crisis. Barriers made a striking comeback, particularly during the first wave of the epidemic in Europe, severely restricting crossings.

These gateways mark successive stages that introduce specific temporalities in the mobility of travelers and goods, providing sovereign authorities with opportunities to make the border clear. The border takes extremely diverse shapes, whether through passport checks, baggage inspections or duty-free purchases, spawning deeply heterogenous and site-specific experiences for passengers (see Chapter 9). These key stages underscore the highly regulated nature of these gateways, relying on extensive control and surveillance apparatuses, applying geopolitically infused mobility regimes to passengers. Depending on their nationality, travel history or destinations, border crossing is either facilitated, complicated or made impossible – restricting the access to specific situations or justifications – and similar rules extend to goods. The Covid-19 health crisis is a case in point, having contributed to the redrawing of traveler access to countries depending on the map of health-related state-to-state relations, according to the rules set by the countries of arrival, transit or departure. These gateways, which are transportation places par excellence, cannot be understood without considering their interdependence, both among themselves and with other categories of gateways.

The interrelationships between gateways are the subject of a recurring debate between two perspectives, particularly regarding site gateways. The first perspective adopts a horizontal interpretation of these connections, viewing them as point-to-point linkages that contribute to a diffuse network of gateways. The route networks of low-cost airlines are a clear example of this dynamic, though some carriers tend to hybridize their model with that of traditional legacy carriers, by adopting greater polarization (Lin and Frétigny 2022). In contrast, the second, more vertical perspective emphasizes the concentration of connections, which reinforces the hierarchy among gateways. Proponents of this view draw on studies that highlight variations in gateway size or more sophisticated rankings. (Air)port networks are envisioned through the hub-and-spokes model (see Chapter 1), where the hub serves as a platform for redistributing flows between other, smaller gateways. Beyond this horizontal and vertical dichotomy, the bipolar organization of European airport hubs gives rise to a more cooperative yet competitive approach – coopetitive – between major airports. These dyads are driven by airline alliances, be they between Amsterdam-Schiphol (KLM's hub) and Paris-Charles-de-Gaulle (Air France's), regarding the Air France-KLM holding, or between London-Heathrow

(British Airways's hub) and Madrid-Barajas (Iberia's) for the OneWorld alliance. In both cases, the hubs tend to share long-haul routes, blurring the representation of gateways as nationally defined entry points.

2.2.2. *Corridor gateways*

Another category, which we have termed corridor gateways, share a similar function with site gateways: facilitating circulation. However, this time, the gateway is distended, operates on a larger scale, and through more streamlined infrastructure. The constant flow of HGVs with a wide range of international license plates on major passageways, such as in Dover and Calais, demonstrates the extent of the internationalization at play through these linear gateways. Suggestively, multiple service areas of highways have been transformed into informal dormitories for international truckers. These gateways become transportation corridors when multiple networks are deployed, multiplying the potential (inter)modal circulation paths. For instance, the Moselle corridor, connecting northern Europe with the Rhone valley combines a variety of routes (highways, railways and waterways) to transport intense flows of goods, cross-border commuters and holidaymakers. Unlike most corridor gateways, which provide international access from one country to another, some very specific gateways, such as the major international straits and the Suez and Panama canals, act as critical intermediaries directly inserted into global exchanges. Although corridor gateways enable both controlled and accelerated flows, they also generate problems, such as environmental nuisances or congestion.

Other corridor gateways take on unexpected configurations, acting as large-scale interfaces for informal migrant crossings: mountains (e.g. the Alps between Italy and France), deserts (e.g. the Sahara and the desertic border zone between the United States and Mexico) and coastlines or seas. Dramatic accounts of shipwrecks in the Mediterranean or the English Channel are piling up, underscoring the harsh realities of these gateways. At another scale, the mass movements of migrants over long distances, often referred to as "caravans", evoke the image of a corridor gateway formed as they pass through.

2.2.3. *Territorial gateways*

They are distinctive for their less ephemeral relationship to living, in that they are recognized residential areas. They are nonetheless linked to an often wide range of economic activities, mobilities and exchanges developed over time. At a large scale, territorial gateways encompass metropolitan regions and cross-border territories, spanning vast contiguous spaces, which are their most emblematic examples. On a finer scale, they include particularly internationalized urban neighborhoods that act as specialized gateways. They comprise districts with a prominent migrant – or even tourist – presence, as well as the more diffuse networks of places that make up diasporic spatialities. These bridgeheads of international relations are also illustrated by embassies and consulates, religious centers, restaurants and specialized shops. Particularly representative of globalization from below, phone and internet cafés play a crucial role, offering diversified services including international money transfers, parcel collection, cell phone sales, etc.

These diasporic gateways blur the dominant representations that emphasize the fragmentation of urban societies. A notable example is in Beirut's affluent neighborhoods, where the phone and internet cafés[2] are frequented by domestic workers, as shown by observations carried out in the Saint-Nicolas residential neighborhood in Achrafieh. These employees, primarily from the Philippines, Sri Lanka and Ethiopia, live with their employers, and these businesses are as close as possible to their place of residence. Another example, countering the stereotype of migrant districts as closed ghettos, is the Burj Hammud district in Beirut's northern suburbs (Figures 2.2(a) and (b)). In fact, this transitional district, accommodating migrants upon their arrival in the city, before they move on to other neighborhoods, combines transportation services, a transportation interchange and other international services. Here, long-standing migrations, such as the Armenian exodus following the genocide of 1913, intersect with more recent migrations of domestic workers, creating a vibrant circulation of people, migrants and non-migrants, and goods, connected to the neighborhood's international bus station, serving routes to Iraq and Syria, and phone and internet cafés, processing parcels to many destinations. Burj Hammud also illustrates the interdependence between site gateways and territorial gateways, as the international bus station is deeply entangled with this diasporic and trading district.

Site gateways and territorial gateways are embedded in globalized interdependences, in which international mobilities play a key role. These linkages are particularly visible in the case of the world city network, where urban societies transcend traditional state-to-state relations. Pandemics precisely raise the issue of interdependent gateways shaped by the mobilities that pass through them. In February 2020, the Covid-19 spread from Chinese metropolises to the other components of the world city network and led to travel restrictions such as direct flight bans from China to Europe and to the United States. They proved too late in hindsight (Kraemer et al. 2020) and partly ineffective in curbing the spread, as other

---

2 Called "taxiphones" locally. Despite the rise of smartphones in the late 2000s, taxiphones are still firmly rooted in the urban landscape.

flights and gateways were used to circumvent the bans, underlining the paradoxical resilience of the network formed by global gateways.

a)

b)

Figure 2.2. *Movement of people, goods and cosmopolitanism in Dora, near Beirut, in the Burj Hammud district (source: Cattan (2012))*

2.2.4. *Embodied gateways*

At a finer scale, what we term embodied gateways are not only gatekeepers but acting as mediators, facilitating various forms of international mobilities. Research on these people is still emerging and this section outlines some of their key features. These embodied gateways can operate outside formal regulations or within most established circuits. Smugglers, for example, involved in the informal crossings of migrants or goods carry out a – more or less suited – logistical mediation, through the transportation equipment they make available or mobilize or the information they provide. Though under different circumstances, sea rescuers also act as embodied gateways. The "frontline rescuer", positioned at the bow of the lifeboat, regulates the passage of migrants through the use of his own body. This individual conveys, in very explicit terms, to use the vocabulary of the lifeguards, whether "the gate" is open or closed, that is, whether boarding is possible, to avoid any panic that might capsize the vessel or cause falls overboard (Banos 2021) (Figure 2.3).

Figure 2.3. *Preparing to rescue migrants in the Mediterranean Sea by the frontline rescuer of the NGO Sea-Watch (source: Grodotzki (2020)). For a color version of this figure, see www.iste.co.uk/fretigny/transportation.zip*

In the very different context of tourism, other actors, known as greeters, engage in territorial mediation: they allow tourists to access more authentic places, practiced by locals. Finally, delivery drivers working for Amazon and other digital platforms are another example of embodied gateways. As they connect the domestic space with global transportation networks, they ensure the globalization of cities at their very doorsteps.

2.3. Why do gateways matter?

Gateways contribute to the construction of regional, national and international territorial systems, raising key questions on territorial dynamics. First, we can think of the role played by networks and hierarchies in the attractiveness of gateways and their capacity to thrive by exploiting the flows that pass through them. Second, the sustainability of these gateways is key to limit their vulnerability, at the forefront of globalization, changes and risks of all kinds. Regulation is the third issue at stake in this construction of territorial systems, as gateways give rise to a multiscale, multidimensional and multi-actor network of interfaces, to manage both inter-territorial and intra-territorial relationships.

2.3.1. *Major attractors: how to endure?*

Attractiveness refers to the material and symbolic influence of the gateway. The gateway and its stakeholders must generate, maintain and renew flows; in doing so, they pay particular attention to the selective and structuring nature of these movements.

Stakeholders value each type of flow differently, establishing hierarchies that underpin the competitiveness of gateways. At least three value registers warrant examination. The first concerns exchange value, where flows are prioritized based on their nature, added value and direction of circulation, rather than volume alone. For example, air freight represents less than 1% of goods traffic by volume, but nearly 35% by value (IATA 2021). Another example is the asymmetric flows of cross-border workers, which for instance can result in prioritizing access for French cross-border workers into Switzerland despite Covid-19, due to their importance to Geneva's economy relatively to other forms of human mobilities. But the value of a flow can also be indirect, through the combination of movements, constituting a second value register. Transit flows, for example, stimulate additional flows in the mobility system. Dubai's authorities take full advantage of them, inviting connecting air passengers to visit the city, in order to boost tourism in the emirate. Finally, a third value register pertains to economic value associated with representations, values and imagination, which are more fundamentally responsible for flow demand. Religious practices and beliefs, for example, and pilgrimages continue to make

Lourdes a specialized gateway into France, supported by a highly active railway station.

The ability of gateways to be inserted into exchanges at various scales also underpins their attractiveness. This integration is based on their accessibility, in terms of time-distance, cost-distance, access speed or ease of travel. But it also depends on the identification and media visibility of gateways themselves and of the places they serve, through the dissemination of positive representations. Many tax havens, for instance, play up their tourist image to consolidate their status as gateways to capital. Short weekend trips or city breaks have flourished not only because of low-cost airlines and targeted tourist products, but also because of shifts in the perception of urban destinations.

Athens, once a one-day summer stopover for international tourists on their way to their final destination, has become a year-round place to stay after the 2008 financial crisis. This transformation is driven by the spread of a specific image, conveyed, by the media, of a resilient urban society, thriving through local solidarity initiatives and collectives and creating an alternative capital of the arts epitomized by street art. This model of a gateway city relies on promoting tourism in working-class neighborhoods, such as Exárcheia, giving rise to significant social tensions. They stem from the proliferation of Airbnb rentals for tourists, and even nomadic workers, making housing less affordable, as well as from the increasing specialization of local shops in tourist sales. Noise pollution, due to suitcase wheels on pedestrian streets and busy cafés and bars, is another factor of friction (Pettas et al. 2022).

The adaptability and innovation of gateways is a final factor in their attractiveness. A significant challenge lies in acknowledging the importance of not only technical innovation but also – and perhaps above all – social and organizational innovation. The logic of technical progress aimed at accelerating air and rail speed has reached its limits, and the challenge of optimizing travel time has largely shifted to gateways. The temptation is high for transportation operators to frame these considerations as a technological issue, focusing on the development of mobile applications, QR codes or RFID chips, to streamline passenger and cargo movement. Yet, organizational innovation is critical in addressing the hazards that affect travelers' waiting times and ever-evolving control procedures, be they regarding security, migration, health or customs, in the wake of terrorist attacks, migratory crises, Covid-19 or Brexit. Social innovation is also crucial, requiring an understanding of how travel time can be best experienced and appropriated by passengers, beyond the logic of reducing travel time, allowing for stress relief, entertainment, work, etc. Social innovation is needed as well for clearer communication and navigation throughout the gateway. The rise of automation is also a key issue, with the proliferation of digital access gates, relying on scanning tickets, passports or fingerprints, reflecting a broader shift of responsibilities from employees to passengers, further exemplified by self-check-in and self-baggage drop-off in air transportation. The considerable complexity of globalized transport networks makes the human presence of agents all the more vital, particularly for communicating in languages familiar to passengers. Interculturality, therefore, represents an important dimension of everyday dynamics of gateways.

2.3.2. *Giants with feet of clay*

Vulnerability, a second major issue for gateways, especially site gateways, refers to anything that affects their sustainability, highlighting their fragility to various risks and changes. The Covid-19 pandemic vividly illustrated gateways' sensitivity to risk, disproportionately affecting the mobility of passengers in site gateways, while spectacularly increasing the visibility and volume of object flows, such as face masks. Corridor gateways continued operating to a large extent, especially for truck traffic and, to a lesser degree, cross-border workers, as in the Moselle corridor. Territorial gateways were also heavily affected by the pandemic, as seen in the dramatic urban exodus of migrant workers in India during the first lockdown, forced onto the roads and through train stations to return to the countryside they came from. Vulnerability also encompasses exposure to environmental and technological risks, exemplified by the 2010 eruptions of the Icelandic volcano Eyjafjöll, which interrupted northern European air traffic for nearly nine days in April and again in May, or the August 2020 explosions at the Beirut port (Figure 2.4). Geopolitics also shapes gateways. For example, Tel Aviv's Ben-Gurion airport rerouted flights to Israel's second-largest airport, Ramon, near Eilat on the Red Sea, following rocket attacks from the Gaza Strip in May 2021.

The vulnerability of gateways also stems from the environmental nuisances they produce, which can undermine their acceptability. Heathrow airport's expansion, Europe's busiest airport in terms of passenger numbers but with only two runways, is emblematic. Despite intense opposition, the British Parliament ultimately approved the project to build a third runway in 2018. Conversely, the same year, plans for a new airport in Notre-Dame-des-Landes, near Nantes, in the middle of a bocage ecosystem – combining pasture, woodland and hedgerows – were abandoned after prolonged occupation of the land by activists labeling it as a "zone to defend". In addition to local environmental nuisances (noise, air pollution, etc.), the footprint of gateways comprises their own contribution to climate change. Current over-consumption of fossil fuels and the international political and legal framework aimed at limiting greenhouse gas emissions contribute to intensifying scrutiny over gateway development projects. The declining acceptability of these projects contrasts with the saturation of traffic in various highly polarizing places and corridors. Touristically, the saturation of site gateways also illustrates the saturation of territorial gateways, of which Venice is particularly emblematic. Cruise ships towering over St. Mark's Square both boost tourism and undermine the city's foundations (Figures 2.5(a) and (b)). They encourage flooding (known as *acqua alta*) and damage the lagoon's ecosystem. This contradiction is undoubtedly one of the primary challenges gateways will face in the coming decades.

Figure 2.4. *The port of Beirut shortly after the explosions of August 4, 2020 (source: Paoli (2020)). For a color version of this figure, see www.iste.co.uk/fretigny/transportation.zip*

Hazard and risks are, more than ever before, integral to societies and their gateways. When it comes to vulnerability, integrating the pace of (climate) changes greatly matters. Knowing how to assess the temporality of risk – how long will the disruption last – and estimating the time required for gateways to return to normal functioning – how long it will take to stave it off – is key for sustainable futures.

a)

b)

Figure 2.5. *Cruise liner in St. Mark's Basin and* acqua alta *phenomenon in the eponymous square in Venice (sources: Dktue (2019) and Alex B (2018))*[3]*. For a color version of this figure, see www.iste.co.uk/fretigny/transportation.zip*

3 Respectively available at: https://commons.wikimedia.org/wiki/File:Panoramic_ photo_ of_a_cruise_liner_in_venice.jpg and https://pixabay.com/photos/venice-st-mark-s-square-flood-water-3459940 [Accessed September 24, 2021].

2.3.3. *Who governs gateways?*

One of the main tensions in the regulation of gateways lies in the shifting power dynamics between public and private actors, constituting the third key issue facing gateways. The place-boundedness of public authorities contrasts with the ability of today's new players to circumvent borders, amid intense power struggles and competition. Indeed, increasingly, it is private operators such as airlines, shipping companies, tour operators and logistics firms that define the importance of gateways, selecting them based on costs and services offered. Maritime shipping companies, for instance, choose port terminals over which they now hold exclusive control. A case in point is Maersk, one of the world's largest container shipowners, through its APM Terminals subsidiary (2024), which has terminals in 60 ports worldwide: Algeciras, Port Said, Tiajin, Los Angeles, Monrovia and more. However, a complex interplay between public and private players can be observed, as in the acquisition of control over many gateway sites by Chinese public and private operators under the Belt and Road Initiative (BRI), with key locations such as Trieste, Piraeus, Gwadar in Pakistan or Mombasa. The rise of private actors also stems from the increasing individualization of flows, which is driven by far-flung mobility of people and goods and exemplified by the rise of e-commerce, generating global flows of freight on a strictly individual scale. These exchanges are not only more numerous, but also far more varied in terms of scale, spaces, modes and durations, fueled by the parallel offerings of low-cost airlines and digital accommodation platforms such as Airbnb.

The question of regulating gateways is even more pressing as they are increasingly disjunct from national borders. The strategic question for gateways is whether they can remain within the reach of public authority, raising the challenge for states of extending their sovereignty over a networked territory. This involves both a shift in representations for public actors and a shift in scale of their competencies. Two examples illustrate this issue particularly well. Since the implementation of Brexit, the United Kingdom and France have faced the challenge of cooperation, made all the more necessary in the broader context of tumultuous negotiations between the European Union and the United Kingdom. Collaboration concerns the Channel Tunnel, to ensure the required border controls while maintaining fluidity on both sides[4], in Dover and Calais. More broadly, it also concerns the Manche interface, marked by the recurrent tragedies of drowning migrants. Beyond bilateral coordination, it is perhaps in the assertion of

4 Considerable queues have regularly formed upstream of Dover, particularly for heavy goods vehicles in December 2020, and for holidaymakers in summer 2022.

transnational institutions, such as the European Union, that public power may regain control over gateways. The question of who, in the years to come, will be responsible for regulating the passage and circulation of people and goods remains unresolved. The example of the Chinese BRI strikingly demonstrates the desire of other state actors to control places of exchange, particularly gateways, well beyond their own borders[5].

The regulation of gateways by public authorities requires them to have the capacity to mediate among a multitude of stakeholders. The aim, for example, is to facilitate intermodality, much sought by operators yet difficult to implement due to the persistent segmentation of gateways by transportation mode. Calls for participatory, bottom-up planning involving civil society present another challenge for public authorities. Acronyms such as NIMBY (Not In My BackYard) or BANANA (Build Absolutely Nothing Anywhere Near Anything) were coined in the United States to describe the opposition of local residents to the construction or extension of facilities, such as waste treatment plants but also highways, airports or high-speed rail links. These complex mobilization phenomena have captivated the imagination of decision-makers, questioning the value given to the rights of individuals and local communities in relation to state and strategic interests.

The use of digital technologies related to the embodied gateways also significantly challenges public action, as it can be endlessly extended, multiplying the places and actors involved. With the rise of e-commerce, spurred by the Covid-19 crisis, the threshold to domestic space becomes internationalized through couriers. The unprecedented connection of these doors to global exchanges raises a key issue of social regulation, regarding working conditions, remuneration and even the migratory status of delivery drivers (see Chapter 4). The concern is also environmental with the proliferation of parcel packaging and the ecological footprint of the associated journeys.

The informal sector in its various forms also presents a regulatory but also ethical challenge for gateways. The diversification of migrant routes and the circumvention of official gateways through smuggling reshape the map of gateways, with the emergence of alternative gateways. For instance, the crossing of the Maroni River between Surinam and French Guiana, for migratory purposes or cocaine trafficking, underscores the importance of this informal corridor gateway. More broadly, Tarrius' (2016) work on "circulatory territories" describes unique, alternative investments in gateways associated with large-scale circulations. He highlights the role of multiple actors in the informal routing of cheap electronic

5 China's initiative prompted in 2021 the European Union's "Global Gateway" and the United States' "Build Back Better World" programs.

products, manufactured in Southeast Asia and purchased in the cities of the Gulf emirates, transported toward the Black Sea and then Western Europe, via the Balkans and the Adriatic. Burgas, the Bulgarian Black Sea port, thus emerges as a key gateway to these circulatory territories. All these practices and circulations, this discreet geography of globalization (Choplin and Pliez 2018) invites us to rethink gateways in the public eye, at street-side, where the informal meets the official. In a very different domain, the illicit trafficking of vegetal and animal lives transiting through these gateways is regularly highlighted by customs discoveries, such as the 185 smuggled turtles found in a passenger's suitcase at the Galapagos Islands airport in March 2021 ((*Le Monde* 2021); for more on these "border encounters" with wildlife in gateways, see Bauer et al. (2019)).

2.4. Conclusion

Gateways are key sites for bringing together distant places and numerous actors, to follow Doreen Massey's (2005) notion of throwntogetherness, thus creating a sense of place through transportation and connections. The routing of exchanges that takes place there makes them key centers for the internationalization of territories and societies. As a result, they are of major interest when it comes to understanding the moving parts and driving forces of globalization as it unfolds. It is perhaps there, more than anywhere else, that the reconnection between here and elsewhere, as well as between centers and peripheries, plays out in all its ambiguity. On the one hand, the concentration and diversity of objects, actors and transportation infrastructures contributes to blur usual assignments. On the other hand, through the borders crisscrossing them, gateways are places where rites of passage are performed, recreating distance and reactivating social and territorial identities. In these places, passage and rootedness, the informal and the official or migrants and tourists get literally close. However, this co-presence of actors and objects does not abolish all the differences of globalization that crystallize there. Multiple categories of mobility are juxtaposed, from smuggled items to luxury objects, from exiles to privileged members of globalized circulation.

On a broader scale, these gateways face strong tensions due to their territorial anchoring and their insertion in networks of interdependent gateways. All planetary issues come into play, including environmental, climate and energy concerns. Internationally, maritime and air transportation – and therefore a significant share of gateways – have seen little regulation regarding greenhouse gas emissions, notably during COP 21, which led to the Paris Agreement. Until now, policymakers have primarily focused on domestic emissions, highlighting the ability of actors associated with these gateways to leverage their international nature to circumvent state commitments. How to ensure an ecologically viable footprint for these gateways remains an open question.

Another source of tension relates to the social and economic issues of these interdependent gateways. The intense competition between gateways involves forms of social dumping, seen in airports in Dubai and Istanbul, or the planned freeport zones in post-Brexit United Kingdom, which may reshuffle the hierarchy of dominant gateways. Tension is also evident at the gateways shaping the borders of the Global North, where they are both significant migratory flows and tight border controls. Ceuta, a Spanish enclave in Morocco, where migrants are repeatedly turned back, exemplifies the humanitarian challenge of gateway regulation. Ceuta is indeed emblematic of the European Union's migration policies projected at its borders. In the coming decades, the challenge will be to design gateway policies where social justice and economic concerns align with environmental and ethical challenges. This all depends on actors' ability to disprove the notion of an impossible sharing of gateways.

2.5. References

APM Terminals (2024). FY 2023 company fact sheet [Online]. Available at: https://www.apmterminals.com/-/media/mainsite/global/corporate/corporate-info/apm-terminals-corporate-factsheet-fy-2023.pdf?rev=7b2e86a14b734e8f93d04da2eaf87f0b [Accessed 30 November 2024].

Banos, A. (2021). Tendre la main : expériences de sauvetage en Mer Égée et en Méditerranée Centrale. In *Séminaire du GT – Mobilités et spatialités*. LabEx DynamiTe, Aubervilliers.

Bauer, S., Güttler, N., Schlünder, M. (2019). Encounters in borderlands: Borderlining animals and technology at Frankfurt Airport. *Environmental Humanities*, 11(2), 247–279. doi: 10.1215/22011919-7754445.

Bird, J. (1983). Gateways: Slow recognition but irresistible rise. *Tijdschrift voor economische en sociale geografie*, 74(3), 196–202.

Burghardt, A.F. (1971). A hypothesis about gateway cities. *Annals of the Association of American Geographers*, 61(2), 269–285.

Cattan, N. (2012). Trans-territoire. Repenser le lieu par les pratiques spatiales de populations en position de minorité. *L'Information géographique*, 76(2), 57–71.

Cattan, N. and Frétigny, J.-B. (2011). Processus et scénarios : les portes d'entrée de la France et les systèmes territoriaux des flux. *Territoires 2040*, 4, 67–84.

Cattan, N. and Frétigny, J.-B. (2021). Street vending in the metropolis: Proximity, distance, and emotions between migrants and tourists in Paris. *The Canadian Geographer/Le Géographe Canadien*, 65(4), 410–422. doi: 10.1111/cag.12700.

Choplin, A. and Pliez, O. (2018). *La mondialisation des pauvres : loin de Wall Street et de Davos*. Le Seuil, Paris.

Frétigny, J.-B. (2016). La conflictualité de la ville en mouvement au regard d'espaces publics métropolitains émergents : les aéroports. In *La ville conflictuelle. Oppositions – Tensions – Négociations*, Desponds, D. and Auclair, E. (eds). Le Manuscrit, Paris.

IATA (2021). Air cargo matters [Online]. Available at: https://www.iata.org/en/programs/cargo/sustainability/benefits/ [Accessed 23 June 2021].

van Klink, H.A. and van den Berg, G.C. (1998). Gateways and intermodalism. *Journal of Transport Geography*, 6(1), 1–9.

Kraemer, M.U.G., Yang, C.-H., Gutierrez, B., Wu, C.-H., Klein, B., Pigott, D.M., Open COVID-19 Data Working Group, du Plessis, L., Faria, N.R., Li, R. et al. (2020). The effect of human mobility and control measures on the COVID-19 epidemic in China. *Science*, 368(6490), 493–497.

Le Monde (2021). 185 tortues découvertes dans une valise à l'aéroport des Galapagos, un policier arrêté. *Le Monde* [Online]. Available at: https://www.lemonde.fr/international/article/2021/03/29/185-tortues-decouvertes-dans-une-valise-a-l-aeroport-des-galapagos_6074812_3210.html.

Lin, W. and Frétigny, J.-B. (eds) (2022). *Low-Cost Aviation. Society, Culture and Environment*. Elsevier, Amsterdam.

Massey, D. (2005). *For Space*. Sage, London.

McKenzie, R.D. (1933). *The Metropolitan Community*. Russel and Russel, New York.

OACI (2021). Rapports annuels du conseil [Online]. Available at: https://www.icao.int/about-icao/Pages/annual-reports.aspx [Accessed 23 June 2021].

Pain, K. (2011). "New worlds" for "old"? Twenty-first-century gateways and corridors: Reflections on a European spatial perspective. *International Journal of Urban and Regional Research*, 35(6), 1154–1174. doi: 10.1111/j.1468-2427.2010.01005.x.

Pettas, D., Avdikos, V., Iliopoulou, E., Karavasili, I. (2022). "Insurrection is not a spectacle": experiencing and contesting touristification in Exarcheia, Athens. *Urban Geography*, 43(7), 948–1006. doi: 10.1080/02723638.2021.1888521.

Prentice, B.E. (2006). Gateways, corridors and strategic city Pairs. In *Competition as a Driver of Change. Proceedings of the 41st Annual Conference of the Canadian Transportation Research Forum*. Université du Saskatchewan, Québec [Online]. Available at: https://mpra.ub.uni-muenchen.de/14625/1/MPRA_paper_14625.pdf [Accessed 23 June 2021].

Rodrigue, J.-P. (2004). Freight, gateways and mega-urban regions: the logistical integration of the Bostwash Corridor. *Tijdschrift voor economische en sociale geografie*, 95(2), 147–161. doi: 10.1111/j.0040-747X.2004.t01-1-00297.x.

Short, J.R., Breitbach, C., Buckman, S., Essex, J. (2000). From world cities to gateway cities. Extending the boundaries of globalization theory. *City*, 4(3), 317–340.

Tarrius, A. (2016). Les routes européennes des nouvelles migrations. Des mobilisations internationales aux mobilités transnationales. In *Conférence au colloque de rentrée du Collège de France. Migrations, réfugiés, exodes*, Paris [Online]. Available at: https://hal.archives-ouvertes.fr/hal-01609941/document.

Ullman, E.L. (1957). *American Commodity Flow*. University of Washington Press, Seattle.

UNCTAD (2021). Review of maritime transport 2020 [Online]. Available at: https://unctad.org/system/files/official-document/rmt2020_en.pdf [Accessed 23 June 2021].

3

Shifting Economic Models of Transportation Places

Juliette MAULAT and Mathilde PEDRO

Géographie-cités, Université Paris 1 Panthéon-Sorbonne, Paris and Aubervilliers, France

3.1. Introduction

In 2020, the Gare du Nord in Paris handled 2,200 trains and 700,000 passengers daily, making it Europe's busiest train station and a critical multimodal hub in the Île-de-France region. Despite significant improvements completed in 2013, the station's ever-increasing passenger volume raises pressing concerns about its capacity limits. In response, SNCF Gares & Connexions, the entity managing France's 3,000 train stations, initiated an ambitious project in 2015 to transform and expand the Gare du Nord. This project, undertaken in partnership with Ceetrus (a subsidiary of Auchan), aimed to create 40,000 m² of commercial and recreational spaces, reorganize the bus station and provide 1,000 bicycles parking space. The project's objectives were manifold: to accommodate the growing passenger flow, alleviate overcrowding on platforms and enhance connectivity between the station and the surrounding neighborhood. Beyond addressing capacity challenges, the project sought to reimagine the Gare du Nord as a "destination" rather than merely a transit point, offering "new services" tailored to passengers, particularly in preparation for the 2024 Olympic Games (SNCF Gares & Connexions 2018). Simultaneously, SNCF Gares & Connexions aimed to generate additional revenue streams to support the station's development, operation and management.

However, the project sparked significant debate following the submission of its building permit in 2019. Initially supported by the Paris City Hall, the project faced

Geographical Places in Transportation,
coordinated by Jean-Baptiste FRÉTIGNY. © ISTE Ltd 2025.

increasing criticism in early 2020 from residents, user groups, citizens[1] and architects (Collectif 2019). Concerns ranged from the project's scale, commercial density and integration with the surrounding neighborhood to the insufficient consideration of daily mobility needs. The involvement of the Auchan group also drew scrutiny, raising fears of privatization and the commodification of public space. Ultimately, in November 2020, the Paris City Hall granted approval for a slightly revised version of the project, which reduced the height of new structures and scaled back retail space. Despite these adjustments, the project was abandoned in September 2021 due to escalating costs and the inability to meet the 2024 deadline. Nevertheless, the Gare du Nord project exemplifies broader transformations in the economic models and management practices of transportation hubs. These transformations are characterized by a shift toward diversified financing mechanisms and increased commercialization, raising controversies about their impact on the development, use and integration of transportation hubs into urban landscapes.

Since the 1990s, the liberalization of transportation markets, changes in public governance and austerity policies have prompted profound shifts in the economic models of transportation infrastructure. These changes are part of global processes of liberalization, privatization and financialization[2] of urban infrastructure (O'Brien et al. 2019a). While these dynamics vary across countries and sectors in Europe, they involve significant restructuring of transportation hubs' economic models. This chapter explores these transformations, examining their causes and implications for the development, use and management of transportation hubs. How did the financing and operation of transportation hubs shift from public models to market-oriented approaches? Here, the term "economic model" encompasses both the methods of financing infrastructure development and the strategies for maintaining and operating transportation facilities. This chapter focuses on stations and airports in order to discuss the transformations in the financing and management of transportation places. They constitute relevant case studies regarding the magnitude of the flows passing through them, their significance and influence in the urban space and the characteristics of the associated mobilities that can involve significant waiting times. While economic models of transportation places are usually addressed at the scale of a transportation sector, this chapter offers to move beyond these segmented approaches by directly addressing different types of transportation places. It focuses primarily on the case of France, brought into perspective with the processes underway on a European and global scale.

1 Including the "Retrouvons le Nord de la Gare" group (http://retrouvonslenord.fr).

2 Financialization refers here to the growing role of the sector, the actors and the logics of finance in public action, in particular infrastructure management (Whiteside 2019).

The chapter begins by examining the shift in financing models, marked by the transition from public funding to increased reliance on private investment and the commercial and real-estate development of transportation hubs. It then explores the mechanisms driving these changes, including (i) the liberalization of the transportation sector, (ii) transformations in the role of the State and (iii) the global circulation of new frameworks, models and financing tools. Finally, the chapter evaluates the implications of these changes for (i) the management and operation of transportation hubs, (ii) their spatial and social integration into urban territories and (iii) the resilience of infrastructure economic models.

3.2. From a public model to a market-oriented model: the evolution of transportation place economics

A long-term perspective highlights the significant transformation of the economic framework governing transportation places. This shift reflects a transition from a traditional public model to a diversified one, which combines public funding with innovative financing mechanisms, such as partnerships with private actors and strategies to enhance the value of transportation spaces through commercial, advertising and real estate activities. These changes have fostered the increasing autonomy of transportation nodes in terms of their operation, management and financing.

3.2.1. *The traditional economic model of stations and airports: network-centric and publicly regulated*

For decades, stations and airports were primarily conceived as network hubs – points of entry, exit or transit within interconnected infrastructure systems. Their financing was historically integrated with that of the broader network, supported by three main revenue streams: public subsidies and borrowing for investment in the creation, expansion and renovation of infrastructure, as well as user fees paid by transportation operators, which covered operating costs. This economic model developed within a context of strict public regulation, which dominated the transportation sector from at least the Second World War[3] until the 1980s for airports and the 1990s for rail infrastructure. During this period, most infrastructure operators were publicly owned and financed, with major national firms such as SNCF (Société

3 The SNCF was founded in 1937 as a semi-public company, with the majority controlled by the French State, replacing public and (above all) private structures. It became a public institution, wholly State-controlled, in 1983. “Aéroport de Paris” (“Paris Airport”) was created as an independent public body in 1945.

Nationale des Chemins de fer Français), Deutsche Bahn (DB) and the British Airports Authority (BAA) overseeing their operation. The natural monopoly of these infrastructures, coupled with their strategic importance for territorial development, led states to consider railway (Finez 2013) and airport (Villard 2009) networks as public goods under tight governmental oversight. For rail networks, public regulation also facilitated mechanisms such as cross-subsidization, which allowed loss-making regional lines to operate using revenue from more profitable ones (Finez 2014).

The 1980s marked the beginning of a shift from a publicly regulated model to one increasingly shaped by market-oriented principles. This transition was driven by multiple factors, including the partial privatization of infrastructure operators, allowing private actors to hold stakes in these firms. However, as Finez (2013) illustrates through the example of SNCF, the distinction between public and private models is often blurred. While SNCF remains fully owned by the French state, it has progressively adopted market-oriented practices, with variations across its activities – such as station management, passenger traffic and freight – depending on the timeline and mechanisms involved. This evolution has been shaped by external and internal forces. Externally, state actors sought to align with the European Union's directives for market liberalization. Internally, shifts in organizational culture and recruitment practices within public enterprises also played a role, as economic and commercial expertise – often brought in from the private sector – gained influence.

These transformations have significantly altered the perception of transportation places. Stations and airports, once seen solely as functional network hubs, are increasingly seen as assets with significant potential for value creation. Two primary resources have been central to these strategies (Jozek 2016): passengers, considered as potential consumers, and the extensive land holdings of infrastructure operators, much of which remains underutilized. This reorientation has contributed to a growing detachment of transportation places from their networked origins, enabling their autonomous management. This shift is reflected in the establishment of specialized subsidiaries and business units dedicated to transportation place management at the turn of the 21st century, such as Grandi Stazioni (1998) in Italy, DB Station&Service (1999) in Germany and SNCF Gares & Connexions (2008) in France. These developments embody the broader trend toward commercialization and professionalization, aligning transportation infrastructure management with market logics.

3.2.2. *New financing instruments: public-private partnerships, commercialization and real estate development*

While public funding and fees related to transportation activities continue to play a significant role in the economic model of airports and stations, financing mechanisms have diversified since the 1990s. This shift occurs within the broader context of evolving infrastructure financing, characterized by a decline in public intervention and an increasing reliance on private and financial actors (O'Brien et al. 2019a). As a result, the investment and operation of transportation places have come to depend more on "alternative" financing instruments (O'Brien et al. 2019b), which can be grouped into several categories. Three major instruments are discussed here: the mobilization of private funding, the enhancement of commercial value and real estate development. One of the primary changes in financing is the growing recourse to private capital, particularly for investment in critical infrastructure, its development, modernization and, in some cases, its operation. This development in Europe has taken various forms, either replacing or complementing public funding, depending on the specific context. Private funding may be mobilized through the sale or establishment of concession contracts for existing transportation places and networks, driven by the limited resources of public entities. In the case of a sale, full ownership of the infrastructure is transferred, while a concession typically involves a long-term lease, with the concession holder responsible for constructing, maintaining and improving the infrastructure. Private funding can also be used for new projects, such as the Tours-Bordeaux high-speed line, which was financed through a public-private partnership. Additionally, private capital may be employed in more intermediate or occasional forms, such as public-private contracts for managing transportation places, where the ownership remains public. A case in point is the 2008 redevelopment of the Paris Gare de l'Est, where the developer, Altarea-Cogedim, contributed 70% of the capital in exchange for a 40-year concession to operate the station's retail spaces (Dang Vu and Jeaneau 2008). A further example is the new bus station in San Francisco, where the Transbay Joint Powers Authority signed a branding agreement with Salesforce. In return for a contribution to the project's financing, Salesforce's name was affixed to the station and the associated office tower for a period of 25 years (Schorung 2019).

Economic models are also increasingly reliant on resources derived from the commercial and advertising development of transportation places. Although this practice is not new (Ribeill 1996, 1999), retail spaces and services within stations have traditionally focused on offering products directly related to travel. However, since the 1980s in the airport sector and more recently in the rail sector, the range of shops and services has expanded to capitalize on passengers' waiting times, as well as to provide a space for urban users to shop and consume. Commercial developments are now often based on agreements between infrastructure managers and commercial entities (Dang Vu and Jeaneau 2008), with managers seeking to leverage the expertise of commercial actors. For instance, Gares & Connexions has partnered with Klépierre, a company that manages numerous shopping centers, while Aéroports de Paris (ADP) works with the Lagardère group, which oversees various press outlets and retail operations within transportation hubs through the Relay brand. These agreements typically involve rental arrangements for commercial leases, with the operator receiving a percentage of the retail turnover. Such travel retail activities now contribute significantly to the revenue of transportation place operators, accounting for more than 30% of ADP's turnover in the Île-de-France region as of 2017 (ADP 2017). Another notable revenue stream is advertising. For example, Transport for London (TfL), which operates the London Underground, signed a 2016 contract to turn stations and rolling stock into advertising spaces, with a projected revenue generation of over £1 billion (O'Brien et al. 2019b).

Beyond commercial development inside stations and airports, real estate development increasingly plays a crucial role in financing transportation places. Rail stations and airport managers are significant landowners. For example, ADP manages 6,686 ha, equivalent to two-thirds of the Paris municipality, while SNCF Gares & Connexions manages nearly 7 million m² of land and 2.7 million m² of station buildings on behalf of the French state. Although transportation operators' real estate activities were initially limited, they have been diversified. Stations and airport' managers now engage in a variety of real estate strategies, such as land transfers, real estate development, and land leasings, etc. These strategies aim to generate short-term revenue through the sale of underused land. For instance, the redevelopment of Toulouse's Matabiau station, EuroSudOuest, is partly financed by the sale of adjacent railway land. Similarly, the Société des Grands Project (2022) responsible for the Grand Paris Express project develops its properties in partnership with private developers. Operators may also pursue longer term strategies, including real estate co-development activities (as seen with TfL) or restructuring as landholding companies (such as SNCF or RATP). This enables them to benefit from capital gains linked to the development, sale and management of real estate assets.

3.2.3. *Differentiated economic models*

The economic models of transportation places are increasingly reliant on revenues derived from "secondary" activities (Magnan 2016), which are not directly tied to transportation operations. For example, in 2017, non-aeronautical activities accounted for 40% of ADP's total revenue, a figure that had been steadily rising over the preceding years (ADP 2017). This growth in revenue from secondary activities is closely linked to the increased use of transportation places, reflecting the "two-sided" nature of these markets (Button 2017). In stations, the rise in user numbers boosts consumption, thereby enhancing the profitability of retail spaces. This, in turn, allows operators to charge higher rents, creating additional resources that can be reinvested in improving station facilities and functionality (Baron 2019).

The growing diversification of financing sources for transportation places has been accompanied by significant shifts in governance and institutional organization. These changes include partnerships with actors specializing in commercial and real-estate value enhancement, the establishment of subsidiaries dedicated to site management and the recruitment of personnel with specialized expertise in these areas (McNeill 2010; Finez 2013). Transportation operators increasingly position themselves as urban development actors. Some, such as the RATP (Delépine 2020), even embrace the concept of the "*ensemblier urbain*" (Lorrain 2002), drawing parallels with private urban service providers (e.g. in water or waste management) and large real-estate developers who have expanded into diversified activities.

However, the degree of diversification varies considerably depending on the country, sector, type and location of infrastructure, traffic levels, local context, operator status and the volume of properties. For instance, airports present significant disparities based on factors such as the extent of their property holdings, air traffic volume, the size of their surrounding metropolitan area and the dynamics of the local real-estate market (Graham 2009; Morrison 2009). At medium-sized airports with limited air traffic and weak real-estate markets, non-aeronautical revenues remain modest, typically accounting for only 10–15% of total turnover, as is the case for Lyon-Saint-Exupéry Airport (Baudoux 2020). Similarly, the reliance on secondary revenues varies markedly among railway stations: regional stations in France remain predominantly funded through public subsidies, whereas major Parisian stations increasingly depend on diversified revenue streams, including retail and advertising.

3.3. Explanatory factors: sectoral restructuring, liberalization, state transformation and circulation of models

What explains the evolution of economic models in transportation places? Three primary factors can be identified: changes in state structures, transformations within the transportation sector and the adoption of new development models.

3.3.1. *Liberalization of transportation and the logics of competition*

The liberalization of transportation markets, aimed at establishing a single European market, is a key factor driving these changes. The pace and implementation of liberalization have varied by country and sector. In aviation, deregulation began in the United States in the early 1970s and unfolded in Europe in three phases. The first two phases (1987 and 1992) eased restrictions on pricing and capacity, while the third phase (1997) removed commercial constraints on airlines, granting them full freedom to operate and set prices within the European Union. Airport charges, however, remain determined at the national level. This competitive environment has prompted airport operators to adjust their strategies to attract both users and airlines. In the rail sector, liberalization came later. European Directive 91/440, adopted in 1991, opened the rail market to competition and mandated the separation of railway infrastructure management from transportation operations. This separation sought to ensure transparent, non-discriminatory access to transportation places and networks for operators, while avoiding competitive distortions caused by unequal public subsidies. In France, European regulatory frameworks – often referred to as "rail packages" – progressively opened up the sector: first, international freight services, then international passenger transportation within the EU (2010), long-distance passenger services (2021) and finally, regional rail services (scheduled for 2024).

The gradual liberalization of the transportation market has had far-reaching implications for the organization of infrastructure operators. In the railway sector, the 1991 directive's requirement for activity separation has been implemented differently across Europe (van de Velde and Röntgen 2009). In France, the SNCF initially adopted an "organizational" separation, creating Réseau Ferré de France (RFF), a state-owned entity dedicated to infrastructure management, in 1997. This was later followed by an "accounting" separation under the 2015 reform, which re-integrated SNCF and RFF into a single public group comprising two distinct branches: SNCF Réseau and SNCF Mobilités, each maintaining separate accounts (Riot 2015).

European competition directives have also transformed the management of stations. In France, Gares & Connexions, originally a subsidiary of SNCF Réseau, gained greater autonomy and became a public limited company in 2020. Liberalization necessitated the financial and governance independence of stations from the broader rail network. This included separate pricing for station access versus rail line access and a distinction between regulated activities (transportation-related operations) and non-regulated activities (such as retail). These changes aimed to ensure transparent pricing. Similarly, in the aviation sector, competition has required a distinction between aeronautical and non-aeronautical activities, with airport fees set transparently for airlines. However, in both sectors, these financial activities are not always entirely independent. This is influenced by the chosen "till" system –whether single-till, dual-till or hybrid models – as illustrated by the case of French stations managed by Gares & Connexions (see Box 3.1).

In France, the 2012 decree established the pricing principles and economic model for Gares & Connexions, drawing on previous SNCF train-station management frameworks[4]. This model is structured around two key elements. First, it categorizes stations based on their size and function, as detailed in the station reference document. The accounting scope of a major station differs significantly from that of a simple stop. Expenses and revenues are allocated to each station or group of stations, collectively referred to as the "management perimeter".

Second, the model distinguishes between "regulated" and "unregulated" domains within each group of stations, adhering to a "dual-till" accounting system. The regulated domain encompasses all in-station activities directly linked to railway operations, such as technical areas, waiting rooms, ticketing services and certain platform spaces like display boards. In contrast, the unregulated domain includes secondary activities, such as retail spaces outside ticketing zones and advertising areas. This distinction serves as the foundation for maintaining two separate accounting systems, hence the term "dual till". Costs and revenues from both domains are consolidated for all stations within the same management perimeter.

For the regulated domain, the fee charged to rail operators for access to station facilities is calculated based on the aggregate costs across all stations in the management perimeter. This results in a uniform rate for the entire perimeter. The station operator is responsible for drafting the pricing plan for regulated services provided within passenger stations, which must be validated by the Transport Regulatory Authority (*Autorité de Régulation des Transports*). For the unregulated domain – such as retail – Gares & Connexions retains full autonomy in setting prices.

However, the two domains are not entirely independent. Since 2003, SNCF's system has allowed for the reinvestment of up to half of the profits generated by unregulated activities to offset regulated costs. This mechanism reduces the fees that rail carriers must pay for station access. For instance, commercial revenues from major Parisian stations are often used to fund station infrastructure improvements, cover a portion of operational costs and lower usage fees charged to railway companies (Cour des Comptes 2021).

Box 3.1 *Regulated and unregulated activities in transportation places: strict separation or interdependence? The example of French stations*

The opening up of competition has been a pivotal factor in driving the diversification of activities among transportation-place operators. This shift has enabled these entities to generate additional revenue streams, invest in infrastructure, enhance service offerings and bolster what is often described as their "competitiveness". For airports, the expansion of retail spaces within terminals and

4 Decree No. 2012-70 of January 20, 2012 relating to rail network service facilities.

the launch of real-estate projects at major hubs since the 1990s are directly tied to the liberalization of the aviation market. By increasingly relying on these revenue-generating activities, airports intend to leverage the growing income from retail and real estate to offset the fees charged to airlines and thus to strengthen their attractivity for airlines companies, as seen in the cases of Hamburg (Morrison 2009), London (McNeill 2010) and Paris (Halpern 2011).

3.3.2. *The restructuring of the state, the doctrines of new public management and austerity policies*

A second set of factors influencing the evolution of economic models for transportation places relates to the restructuring of state intervention mechanisms and the changing status of transportation-place management entities. Although urban infrastructure in Europe remains largely controlled and funded by states, the increasing privatization of transportation companies, neoliberal doctrines of public management and austerity policies have driven operators toward diversification strategies. As Adisson (2018) demonstrates in the case of railways, public entities are increasingly obligated to bolster their capacity for self-financing to sustain operational activities and investments alongside public funding.

The spread of new public management doctrines, characterized by an emphasis on "performance", optimization and rationality in the public sector, has accelerated these shifts. France's SNCF provides a notable example: the "performance contracts" between the company and the state stipulate self-financing targets, including leveraging its assets and selling land to reduce its debt (Adisson 2018). Governments are actively encouraging public transportation firms to adopt alternative financing mechanisms, such as public-private partnerships, land-value capture strategies and financialized funding channels. These approaches are exemplified in London's urban transportation system, as analyzed by O'Brien et al. (2019b). Budgetary austerity policies further intensify these pressures. For instance, in the case of Gares & Connexions, while stations continue to benefit from significant public funding and user fees[5], the entity faces a "scissor effect": public funding is stagnating or even declining, while costs – including those for maintenance, operations and infrastructure investments – are rising. This is due to aging infrastructure, increased traffic and the growing demand for modernization and security upgrades. In many national contexts, declining public funding is the primary driver of diversification strategies. In India, for example, the Delhi Metro Rail Corporation has, since the 1990s, actively developed its land and advertising

5 In 2018, investments by Gares & Connexions (SNCF Gares & Connexions n.d.) in stations amounted to more than €380 million, subsidized by around 40%.

assets within transportation hubs to offset drastic reductions in central government funding for urban transportation (Bon 2017).

The corporatization of public transportation firms has reinforced these dynamics, particularly by removing legal barriers tied to their public status. For example, the transformation of ADP into a public limited company in 2005 was accompanied by the transfer of land ownership from the state to the operator and the removal of restrictions associated with the "specialty" legal principle, thereby permitting real-estate development activities (Halpern 2011). Similar processes are evident in Japan, where the privatization of Japan National Railways (JR) in 1987 significantly spurred the growth of secondary activities (Aveline 2003). The partial or full privatization of transportation operators, coupled with the increasing financialization of their activities, has also reinforced diversification trends. Privatized airports, for instance, diversify not only to fund terminal operations and development but also to generate higher returns for shareholders. Land and real-estate assets often play a pivotal role in attracting private investors, as evidenced in the cases of London-Heathrow (McNeill 2010) and Brussels-National (Deruytter and Derudder 2019) airports. Transportation places are thus becoming increasingly financialized, both through the growing influence of financial actors and the "discreet" adoption of financial logics and tools by their management, as noted by Lorrain (2011).

3.3.3. *The circulation of new urban planning and financing models*

The evolution of transportation-place financing can also be attributed to the global circulation of financing tools, management frameworks and urban models. For instance, the Japanese station model, which leverages real-estate value enhancement and draws on practices pioneered by early American tram companies, has served as a significant source of inspiration for managing European stations (Aveline 2003; Tiry-Ono 2018). These international exchanges are often cross-sectoral: the SNCF, for example, frequently cites ADP and, more broadly, the airport model as key references for station management and redevelopment (Dang Vu and Jeaneau 2008; Baron 2019). Such exchanges are driven by international capital providers, such as the World Bank, which promote these financing models. They are also facilitated by industry gatherings, such as the International Union of Railways (UIC), which organizes multi-year events like the Next Station conference and hosts a dedicated working group on stations (Riot 2015). Additionally, these circulations are supported by consulting firms, including subsidiaries of ADP or Gares & Connexions, and architectural agencies like AREP. Strategic partnerships between infrastructure operators, such as the collaboration between NS (the Dutch railway company) and JR Kyushu in Japan, further exemplify these dynamics (van de Velde 2011).

The adoption of these new economic models for transportation places is closely aligned with prevailing urban-planning paradigms that emphasize urban densification, renewal and the integration of urban planning with transportation networks. For example, the "airport city" model, popularized by Kasarda (2006), has been employed by operators to legitimize diversification strategies. In rail transportation, similar diversification efforts are rooted in the transit-oriented development (TOD) model, which promotes urban densification around stations. This approach significantly shapes regional planning across Europe, North America and Asia (Bertolini et al. 2012). These urban models serve as tools to justify real-estate value-enhancement initiatives that help finance infrastructure, as studied in Europe (Gallez et al. 2015), the United States (Schorung 2019) and Asia (Bon 2017).

Local public actors often play a crucial role in driving diversification processes. The commercialization of station spaces, real-estate developments around stations or the establishment of office zones on airport platforms are frequently seen by local governments as opportunities to enhance neighborhood value, improve infrastructure integration, address public policy objectives (e.g. employment, amenities and housing) and generate new tax revenues. Such developments align with broader urban growth strategies and are often supported by "growth coalitions" that bring together infrastructure operators, real-estate developers and local governments. The Crossrail project in London is a prominent example of this collaborative approach (Mboumoua 2017; O'Brien et al. 2019b). Nevertheless, local governments sometimes oppose such projects when they conflict with urban policies, face resistance from residents or businesses or result in negative externalities. For example, the Stuttgart 21 rail project in Germany sparked significant opposition due to its perceived contradictions with local interests (Novy and Peters 2012; Volin 2020). Similarly, real-estate operations around Sydney and Perth airports have highlighted governance challenges, with local authorities often holding limited influence relative to the significant powers of infrastructure operators (Morrison 2009). But such conflicts remain relatively rare. Research more frequently highlights the active support of local public actors for commercialization or real-estate development at transportation places.

3.4. Effects of these changes on the functions, forms and practices of stations and airports: models in question

The transformation of financing, investment and operational strategies for transportation hubs influences their development, urban integration and the practices they shape in terms of mobility and functionality. These shifts invite critical reflection on the alignment of these new models with the core purposes of transportation hubs: facilitating mobility, circulation and transit.

3.4.1. *The complex interplay between transit and commercial functions*

While the primary role of transportation hubs is to facilitate passengers' access to networks, the development of stations and airports increasingly reflects the coexistence of their transit and mobility functions with commercial objectives aimed at monetizing passenger and user flows.

During the 1980s and 1990s, when the first multimodal interchange hubs in France were designed to combine transit and commercial functions, a hierarchical logic prevailed. The circulation of passengers was prioritized, with services directly related to transportation positioned as close as possible to passenger flows, while ancillary services were relegated to peripheral or parallel zones to avoid impeding mobility (Gille 1999). Today, this prioritization has shifted: transportation-place operators are increasingly focused on capturing passenger flows (Baron 2019), redirecting them toward shops and services to maximize secondary revenue, in contrast to the earlier emphasis on facilitating rapid access to transportation. For instance, at London's St. Pancras International station, the complex spatial configuration, once considered a challenge for passenger circulation, is now leveraged as an asset in the station's commercial strategy. The absence of a centralized waiting area adjacent to the platforms and the necessity for passengers to traverse multiple levels enhances the visibility of retail spaces, which are strategically located along passenger pathways (Riot 2015).

Similarly, in airports, the design and expansion of retail areas – particularly duty-free shops – reflect similar flow-capture strategies (Figure 3.1). These strategies are even more pronounced given passengers' longer dwell times in airports compared to stations. By positioning retail spaces near checkpoints, operators create a deliberate contrast between restrictive security procedures and areas intended for relaxation and consumption (Frétigny 2013). This positioning not only facilitates consumption but often makes retail spaces an unavoidable passageway en route to boarding gates. An illustrative example of such strategies is Oslo-Gardermoen Airport, as depicted in maps by Rekacewicz (2013) for *Le Monde diplomatique*. In 2005, the duty-free shop was located in a peripheral section of the international terminal. By 2006, it had been repositioned as a corridor directly opposite security checkpoints. In 2007, retail spaces expanded further, reducing waiting areas and forcing passengers to traverse the commercial zone before reaching boarding gates, even at the cost of impeding mobility. Transportation hubs are thus increasingly reimagined and redeveloped not only to address passengers' expectations but also to maximize their commercial and advertising potential. Entire spaces within hubs are monetized. Grandi Stazioni, the operator of Italy's international train stations, explicitly markets these stations as "settings for large-scale advertising operations" (Riot 2010, p. 6). The economic model of these hubs profoundly influences their visual landscape, with a proliferation of shop signage, announcements and displays. However, this commercial emphasis often conflicts with the logic of prioritizing navigational information essential for passenger orientation.

Figure 3.1. *A retail space installed directly along the flows at Amsterdam-Schiphol Airport (source: Frétigny (2012)). For a color version of this figure, see www.iste.co.uk/fretigny/transportation.zip*

The transformation of the economic models of stations and airports also accentuates their role as selective and stratified spaces. While the cost of transportation itself has often decreased with the advent of low-cost travel options, the proliferation of new services sustains and reinforces social hierarchies. Premium services, such as fast-track access to border checkpoints (Frétigny 2013) or exclusive lounges, are now integral to the economic strategies of airport and, increasingly, station operators. This growing stratification can alienate certain users, particularly those excluded from the targeted consumer base of commercial and advertising campaigns. In airports, for example, advertising frequently focuses on luxury products and reinforces the perception of air travel as a privilege of affluent social classes (Frétigny 2015). Passengers outside these classes may feel their legitimacy to occupy and navigate airport spaces is undermined. Finally, transportation hubs, traditionally seen as refuges for marginalized populations, such as homeless individuals, are increasingly inaccessible as these spaces are repurposed for retail and commercial use. This exclusion reflects broader societal shifts in the economic and social functions of transportation places (Domingo 2007).

3.4.2. *The territorial consequences of transforming transportation places: densification, competition and citizen opposition*

This evolution also impacts the territorial and urban integration of transportation hubs, influencing their relationship with the cities. Airports, and especially train stations, have become focal points for urban densification policies, driven by a convergence between the land development strategies of infrastructure operators and public policies aimed at sustainable urban development, notably through the model of TOD. Public authorities are increasingly seeking to develop underutilized properties of infrastructure operators for urban and social policies. For example, on May 28, 2021, the SNCF and the French Ministry of Housing signed a charter committing to mobilizing public railway properties for housing development. This agreement includes transferring rights of way to build between 15,000 and 20,000 housing units, with 35% designated as social housing (Ministère de la Transition écologique 2021). Similarly, in Italy, the Municipality of Milan authorized the state-owned railway company, Ferrovie dello Stato, to undertake land value-enhancement projects, negotiating in return for the creation of green spaces and social housing (20–35%) as well as reinvesting the capital gains generated into the improvement of the Milan railway hub (Riot and Adisson 2011).

However, the growing interest in these new urban models and the increasing demand for land, real estate, commercial and advertising value enhancement by infrastructure operators present several challenges regarding the role of transportation hubs and networks in shaping urban space. A key issue is the proliferation of large-scale projects around transportation hubs in French and European metropolitan areas (Delage 2018), which contributes to the standardization of metropolitan landscapes and a potential oversupply, particularly in tertiary areas, which are often deemed more profitable by infrastructure operators. The Grand Paris Express, for example, encompasses over 350 development projects around its 68 new stations, covering a total of 32 million m^2, including 9.2 million m^2 of commercial and office space. These developments indirectly finance the network through the Ile-de-France tax on office, commercial, storage, and parking spaces (TSBCS), benefiting the Société des Grands Projects. These initiatives are generally viewed positively by stakeholders of the new network, but they often overlook the broader urban effects associated with such large-scale real-estate development.

a)

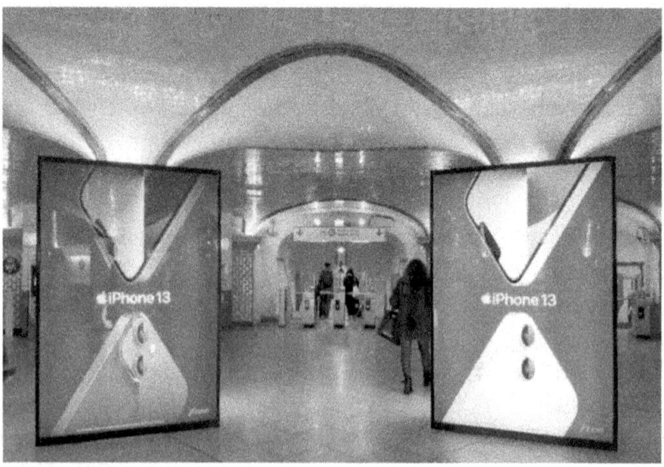

b)

Figure 3.2. *Shopping malls (SNCF space) and advertising campaign (RATP space), Saint-Lazare station in Paris (source: Pedro (2021)). For a color version of this figure, see www.iste.co.uk/fretigny/transportation.zip*

Moreover, the shift in the economic model of transportation hubs toward a focus on commercial and service functions raises concerns, particularly regarding the competition it may create with existing local businesses. Retail spaces within transportation hubs can be seen negatively by local stakeholders, who fear an imbalance in the market. This was evident in the case of Aéroville, a shopping center developed by ADP in 2013, featuring 200 shops near Roissy Airport (Fischer 2020). Its construction in an area already saturated with large retail spaces led to opposition from local traders and elected officials, who expressed concerns about the potential negative impact on city-center businesses. Similarly, Dang Vu and Jeaneau (2008) highlight commercial competition at Paris' Saint-Lazare station. After a major renovation in 2012, the station now houses a 10,000 m² shopping mall. However, it faces stiff competition from the surrounding neighborhood, one of the densest retail areas in Europe. The challenge for SNCF and Ségécé (Klépierre's subsidiary managing the station's shopping center) is to design the station in such a way as to encourage passengers to make purchases before leaving (Figure 3.2(a)). This strategy runs counter to the vision of the station as a "piece of the city", integrated into its environment, a vision supported by AREP and the Paris municipality elected officials.

Stuttgart 21 is an ambitious urban development project led by Deutsche Bahn (DB) around Stuttgart's central station. Launched in the 1990s, the project aims to modernize the rail hub as part of the Paris-Budapest high-speed rail corridor (Magistrale for Europe). It involves replacing the aboveground terminal station with an underground through station, thereby freeing nearly 100 ha of valuable surface area in the city center for redevelopment (Volin 2020). The project's initial cost estimate of €2.5 billion was intended to be largely offset by the land value enhancement of the DB-owned surface area.

While supported by the railway company and local officials – who viewed it as an opportunity to reduce the urban footprint of rail infrastructure and stimulate economic development – the project quickly became a focal point for widespread citizen opposition. Several factors fueled this dissent, including concerns about the project's economic viability and escalating costs, minimal transportation benefits, significant environmental impacts and urban planning decisions perceived as prioritizing financial interests over social, environmental or architectural considerations. Furthermore, the opacity of decision-making processes and the lack of public consultation aggravated tensions.

The resulting protests were unprecedented in Germany, both in scale and in the diversity of participants (Novy and Peters 2012). A mediation process and a local referendum held in 2011 ultimately allowed Stuttgart 21 to proceed, but it remains emblematic of resistance to railway megaprojects. The project's core tensions stemmed from its prioritization of railway company interests over citizens' needs, its questionable democratic legitimacy and the perception that real-estate development overshadowed the infrastructure's intended purpose.

Box 3.2. *Stuttgart 21, an example of local opposition to a railway megaproject around the city's central station*

Lastly, the pursuit of revenue diversification by transportation hub operators significantly impacts land and real-estate prices. The proximity to efficient transportation infrastructure, coupled with strategies favoring the most profitable activities – such as high-end residential developments, office spaces and shopping centers – contribute to inflation in land values. Aveline-Dubach (2023) notes that it is unsurprising that countries or regions adopting land-based financing instruments (e.g. Japan, Hong Kong and China) are now among the world's most expensive residential real-estate markets. In Hong Kong, civil society has increasingly opposed the "rail + property" model employed by the Mass Transit Railway Corporation (MTRC), which owns the subway network. While this model successfully finances the construction and operation of the network, it exacerbates rising property prices, social stratification and extreme neighborhood density. This example highlights the growing citizen opposition to projects perceived as prioritizing land value capture over local needs. A prominent illustration is the Stuttgart 21 project, which has sparked significant controversy and public dissent (see Box 3.2).

3.4.3. *Are economic models for transportation places weakening?*

The robustness of the new economic models implemented by transportation place operators is increasingly under scrutiny. These models are inherently dependent on the urban, land and institutional context of the transportation hubs in question. Secondary revenues rely heavily on projected growth in passenger flows: the anticipated increase in passenger numbers is often cited to justify renovation projects for rail and airport terminals, which are designed to maximize commercial value. Flow volume and proximity also serve as key indicators for setting commercial lease rates, as they directly impact the profitability of shops by increasing the "stop ratio" (Detavernier 2018), the percentage of passersby who interrupt their journey to visit the shops. While these economic models have demonstrated financial viability in large metropolitan stations and international airports, the same cannot always be said for smaller hubs. Roudier (2019) notes that SNCF Gares & Connexions struggles to achieve financial balance in stations located in medium-sized cities. In such cases, the cost of renting station spaces often exceeds the financial capacities of local actors. This has led to the emergence of a two-speed network: profitable metropolitan transportation hubs that attract private investment and secondary hubs essential for regional connectivity but reliant on limited public funding, which is frequently contested.

The stability of infrastructure financing based on land and real-estate value enhancement is also under scrutiny. O'Brien et al. (2019b) underscore the speculative nature of these revenues, making them highly sensitive to fluctuations in real-estate markets as well as changes in institutional frameworks and urban-planning regulations. For example, TfL's real-estate ambitions were curtailed by a downturn in the city's high-end housing market (O'Brien et al. 2019b). Similarly, in the Netherlands, the post-2008 economic crisis collapse of real-estate markets forced municipalities and train station operators to rethink their models for financing railway hub renovations. The reliance on real-estate value enhancement became unsustainable, leading to a strategic shift toward separating the funding of transportation infrastructure from urban planning initiatives (Janssen 2021). Although Asian models for financing infrastructure through land and real-estate value enhancement are often lauded, their success is deeply rooted in specific urban, institutional and political contexts. Aveline-Dubach (2023) highlights that these models are now facing challenges due to evolving circumstances. In Japan, railway companies' land value capture strategies, originally applied at the scale of entire rail lines, are losing relevance as urban shrinkage, driven by aging populations and peripheral socioeconomic decline, takes hold. Declining passenger traffic, falling real-estate values and increasing property vacancies have compelled railway companies to focus on major stations and asset management to maintain long-term revenue. Ironically, these real estate development strategies, initially intended to bolster companies' revenues, have exposed them to greater vulnerability in the face of urban and demographic challenges.

The financial viability of these models hinges not only on passenger flow growth but also on broader urban, demographic and economic trends. As such, they are highly sensitive to external shocks. The Covid-19 pandemic starkly illustrated this fragility, leading to plummeting transportation place footfall, shop closures and significant disruptions in real-estate markets.

3.5. Conclusion

The economic models of transportation places have undergone significant transformation since the 1980s, reflecting broader global trends in urban infrastructure and shifts in state action. The comparison between stations and airports reveals a convergence of strategies among infrastructure operators and a cross-sector circulation of models, despite sector-specific differences. These changes go beyond mere financial restructuring; they influence the development and management of transportation spaces and, by extension, the experiences of mobility and user practices. However, these evolving models remain underexplored. Further research is needed to examine the financing of transportation places and their broader implications, whether for train stations, airports or ports, as well as for less-documented but essential spaces such as bus stations.

3.6. References

Adisson, F. (2018). From state restructuring to urban restructuring: The intermediation of public landownership in urban development projects in France. *European Urban and Regional Studies*, 25(4), 373–390.

ADP (2017). Rapport d'activité. Aéroports de Paris, Tremblay-en-France.

Alterman, R. (2012). Land use regulations and property values: The "Windfalls Capture" idea revisited. In *The Oxford Handbook on Urban Economics and Planning*, Brooks, N., Donaghy, K., Knaap, G.-J. (eds). Oxford University Press, Oxford.

Ashton, P., Doussard, M., Weber, R. (2016). Reconstituting the state: City powers and exposures in Chicago's infrastructure leases. *Urban Studies*, 53(7), 1384–1400.

Aveline, N. (2003). *La ville et le rail au Japon : l'expansion des groupes ferroviaires privés à Tôkyô et Ôsaka*. CNRS Editions, Paris.

Aveline-Dubach, N. (ed.) (2023). Land value capture and its large-scale application in northeast Asia. In *Globalization and Dynamics of Urban Production*. ISTE Ltd, London, and John Wiley and Sons, New York.

Baron, N. (2019). Designing Paris Gare du Nord for pedestrians or for clients? New retail patterns as flow optimization strategies. *European Planning Studies*, 27(3), 618–637.

Baudoux, M. (2020). Les stratégies de valorisation foncières et immobilières de l'aéro-port Lyon Saint-Exupéry. Master's Thesis, Université Paris I, Paris.

Bertolini, L., Curtis, C., Renne, J. (2012). Station area projects in Europe and beyond: Towards transit oriented development? *Built Environment*, 38(1), 31–50.

Bon, B. (2017). Le rôle des transports collectifs dans l'aménagement des régions métropolitaines. Le "Transit Oriented Development" et la construction du métro de Delhi. *Bulletin de l'association de géographes français*, 94(1), 62–76.

Bon, B. (2021). Making railway land productive: The commodification of public land in Kenyan and Indian cities. *Geoforum*, 122, 118–128.

Button, K. (2017). The other side of an airport's two-sided market: Issues in planning and pricing airport surface access. *European Journal of Transport and Infrastructure Research*, 17(4). doi: 10.18757/ejtir.2017.17.4.3208.

Collectif (2019). Le projet de transformation de la gare du Nord est inacceptable. *Le Monde* [Online]. Available at: https://www.lemonde.fr/idees/article/2019/09/03/le-projet-de-transformation-de-la-gare-du-nord-est-inacceptable_5505639_3232.html [Accessed 20 September 2021].

Cour des comptes (2017). La société du Grand Paris. Report, Paris [Online]. Available at: https://www.ccomptes.fr/sites/default/files/2018-01/20170117-rapport-societe-grand-Paris_1.pdf [Accessed 30 August 2021].

Cour des comptes (2021). Les gares ferroviaires de voyageurs. Report, Paris [Online]. Available at: https://www.ccomptes.fr/sites/default/files/2021-04/20210420-rapport-gares-ferroviaires-voyageurs.pdf [Accessed 30 August 2021].

Dang Vu, H. and Jeaneau, H. (2008). Concevoir un espace de transit et de consommation : la gestion de site dans les gares parisiennes. *Espaces et sociétés*, 135(4), 45–62.

Delage, A. (2018). Gare au standard ! Les nouveaux quartiers de gare TGV, produit d'appel pour des territoires en mal de reconnaissance ? *Les Annales de la recherche urbaine*, 113(1), 150–165.

Delépine, I. (2020). Les stratégies de valorisation foncière et immobilière de la RATP : le cas des "sites industriels". Master's Thesis, Université Paris I, Paris.

Deruytter, L. and Derudder, B. (2019). Keeping financialisation under the radar: Brussels Airport, Macquarie Bank and the Belgian politics of privatised infrastructure. *Urban Studies*, 56(7), 1347–1367.

Detavernier, P. (2019). La gare est-elle un centre commercial comme les autres ? *Réflexions immobilières*, 89, 67–72.

Domingo, B. (2007). "SDF" et construction d'un ordre public local : fluidités de l'identité assignée et normalisation des lieux. *Déviance et société*, 31(3), 283–303.

Finez, J. (2013). Aux origines de la préoccupation marchande à la SNCF (1960-2011). *La nouvelle revue du travail*, 2. doi: 10.4000/nrt.909.

Finez, J. (2014). La construction des prix à la SNCF, une socio-histoire de la tarification. *Revue francaise de sociologie*, 55(1), 5–39.

Fischer, A. (2019). Les stratégies de valorisation foncière et immobilière d'ADP. Master's Thesis, Université Paris 1, Paris.

Frétigny, J.-B. (2013). La frontière à l'épreuve des mobilités aériennes : étude de l'aéroport de Paris Charles-de-Gaulle. *Annales de géographie*, 690(2), 151–174.

Frétigny, J.-B. (2015). La catégorie de mobilité naturalisée par les lieux de passage : étude de la publicité à l'aéroport de Roissy Charles-de-Gaulle. *Regards sociologiques*, 1(45/46), 47–69.

Gallez, C., Maulat, J., Roy-Baillargeon, O., Thébert, M. (2015). Le rôle des outils de coordination urbanisme-transports collectifs dans la fabrique politique urbaine. *Flux*, 101/102(2), 5–15.

Gille, L. (1999). Du rapport entre pôle et place d'échanges, In *Villes en gares*, Joseph, I. (ed.). Éditions de l'Aube, La Tour d'Aigues.

Graham, A. (2009). How important are commercial revenues to today's airports? *Journal of Air Transport Management*, 15(3), 106–111.

Graham, A. (2019). Airport management: A perspective article. *Tourism Review*, 75(1), 102–108.

Halpern, C. (2011). Portrait d'entreprise. HUB 4. Bâtisseur, gestionnaire et aménageur : les stratégies adaptatives de Fraport AG et Aéroports de Paris. *Flux*, 84(2), 73–89.

Janssen, C. (2019). Spoorzone Delft : quatre leçons pour la densification urbaine dans les quartiers de gare [Online]. Available at: https://www.gebiedsontwikkeling.nu/artikelen/spoorzone-delft-4-lessen-voor-stedelijke-verdichting-stationsgebieden/ [Accessed 30 August 2021].

Jozek, F. (2016). Les ressources extra-aéronautiques des aéroports commerciaux. *Droit et ville*, 82(2), 3–18.

Kasarda, J.D. (2006). Airport cities and the aerotropolis [Online]. Available at: http://aerotropolis.com/airportcity/wp-content/uploads/2018/10/2006_07_AirportCitiesAndTheAerotropolis-1.pdf [Accessed 31 August 2021].

Lorrain, D. (2002). Capitalismes urbains : la montée des firmes d'infrastructures. *Entreprises et histoire*, 30(3), 7–31.

Lorrain, D. (2011). La main discrète. *Revue française de science politique*, 61(6), 1097–1122.

Magnan, M. (2016). La production et la gestion de l'espace portuaire à vocation industrielle et logistique. Les grands ports maritimes français : gestionnaires d'espaces infrastructurels. PhD Thesis, Université Paris I Panthéon-Sorbonne

Mboumoua, I. (2017). Revisiting the growth coalition concept to analyse the success of the Crossrail London megaproject. *European Planning Studies*, 25(2), 314–331.

McNeill, D. (2010). Behind the "Heathrow hassle": A political and cultural economy of the privatized airport. *Environment and Planning A*, 42(12), 2859–2873.

Ministère de la Transition écologique (2021). Relance de la production de logements : Emmanuel Wargon et Jean-Pierre Farandou signent une charte pour la mobilisation du foncier ferroviaire [Online]. Available at: https://www.ecologie.gouv.fr/relance-production-logements-emmanuel-wargon-et-jean-pierre-farandou-signent-charte-mobilisation-du [Accessed 30 August 2021].

Morrison, W. (2009). Real estate, factory outlets and bricks: A note on non-aeronautical activities at commercial airports. *Journal of Air Transport Management*, 15, 112–115.

Musil, C. and Perset, M. (2015). Le financement des infrastructures de transport à Hô Chi Minh-Ville (Viêt-Nam) : les outils, les innovations et les défis. Working Paper, PADDI.

Novy, J. and Peters, D. (2012). Railway station mega-projects as public controversies: The case of Stuttgart 21. *Built Environment*, 38(1), 128–145.

O'Brien, P., O'Neill, P., Pike, A. (2019a). Funding, financing and governing urban infrastructures. *Urban Studies*, 56(7), 1291–1303.

O'Brien, P., Pike, A., Tomaney, J. (2019b). Governing the "ungovernable"? Financialisation and the governance of transport infrastructure in the London (global city-region). *Progress in Planning*, 132, 100422.

Rekacewicz, P. (2013). L'aéroport à l'assaut des passagers. *Le Monde diplomatique*, February, 13–16.

Ribeill, G. (1996). Les métamorphoses de la grande gare française. *Les Annales de la recherche urbaine*, 71(1), 54–65.

Ribeill, G. (1999). D'un siècle à l'autre, les enjeux récurrents de la gare française. In *Villes en gares*, Joseph, I. (ed.). Éditions de l'Aube, La Tour-d'Aigues.

Riot, E. (ed.) (2010). La gare internationale à l'heure de l'ouverture à la concurrence du rail : jeux d'acteurs et reconfiguration des enjeux. In *École thématique "Controverses. Le marché fait-il la ville ?* CNRS, Aussois [Online]. Available at: https://halshs.archives-ouvertes.fr/halshs-00638052.

Riot, E. (2015). L'agencement des grandes gares historiques pour le marché ferroviaire européen : analyse comparée de l'intégration des principes concurrentiels dans l'aménagement et la gestion des gares de London St Pancras, Paris Nord et Milano Centrale. PhD Thesis, Université Paris-Est, Marne-la-Vallée.

Riot, E. and Adisson, F. (eds) (2011). Paris-Milan : comparer l'intégration européenne dans la production de la ville. Regards croisés sur les questions ferroviaires et d'urbanisme. In *École thématique "Les controverses de l'action publique"*. CNRS, Aussois.

Roudier, E, (2018). Quand la ville moyenne entre en gare : des projets entre mimétisme métropolitain et recompositions territoriales. PhD Thesis, Université Paris-Est, Créteil.

Schorung, M. (2019). New stations and transit-oriented development in three United States rail corridors. In *Transit Oriented Development and Sustainable Cities*, Knowles, R.D. and Ferbrache, F. (eds). Edward Elgar Publishing, Cheltenham, 12–28.

SNCF Gares & Connexions (2018). Bilan de la concertation préalable. Grand projet Paris Gare du Nord 2024 [Online]. Available at: https://www.garesetconnexions.sncf/sites/default/filesd7/field_files/2018-01/synthese_consultation_publique_gare_du_nord.pdf [Accessed 30 August 2021].

SNCF Gares & Connexions (n.d.). Modèle économique [Online]. Available at: https://www.garesetconnexions.sncf/fr/entreprise/entreprise-modele-economique [Accessed 30 August 2021].

Société du Grand Paris (2022). Les quartiers de gare, nouveaux lieux de vie [Online]. Available at: https://www.societedugrandparis.fr/ville-demain/quartiers-gare [Accessed 20 October 2022].

Thomas, A. (2019). Les projets immobiliers connexes de la Société du Grand Paris. Master's Thesis, Université Paris 1, Paris.

Tiry-Ono, C. (2018). *L'architecture des déplacements : gares ferroviaires du Japon*. Infolio, Gollion.

van de Velde, D. (2011). The Japanisation of the Dutch railways. *Journal of Business Studies*, 57(3), 25–57.

van de Velde, D. and Röntgen E.F. (2009). Railway separation European diversity. In *Internationalization of Infrastructures: Proceedings of the 12th Annual International Conference on the Economics of Infrastructures*, Auger, J.F., Bouma, J.J., Kunneke, R. (eds). Delft University of Technology, Delft.

Villard, P. (2009). Aéroports de Paris : du service public à l'entreprise de services : émergence et mise en action d'un nouveau référentiel des politiques aéroportuaires. Master's Thesis, Institut de Sciences Politiques, Paris.

Volin, A. (2020). Des territoires au coeur d'un conflit d'aménagement. Stuttgart 21 et la Magistrale européenne Paris-Budapest. PhD Thesis, Université Lumière Lyon II, Lyon.

Whiteside, H. (2019). Advanced perspectives on financialised urban infrastructures. *Urban Studies*, 56(7), 1477–1484.

4

Logistics Places: An Urban Geography of Post-Industrial Blue-Collar Workers

Nicolas RAIMBAULT
ESO Nantes Université, France

4.1. Introduction

The Covid-19 pandemic and the various lockdown and curfew episodes in 2020 and 2021 cast light on places, flows and workers that had largely remained in the shadow of post-Fordist transformations and neoliberal policies at work in metropolitan areas[1]. Due to shortages, the challenges related to supplying daily consumer goods were widely covered in the media. Delivery drivers found themselves virtually alone in the streets; and order pickers and forklift operators, responsible for handling goods and parcels in warehouses, had to continue to go to work in logistics parks (Benvegnù et al. 2021) while business districts were deserted. In addition, since the late 2010s, social movements around warehouses have multiplied (notably relating to Amazon), such as strikes and demonstrations by digital-platform delivery drivers. Surveys of the *gilets jaunes* ("Yellow Vest")

1 The transition from Fordism to post-Fordism corresponds to the emergence of globalized, flexible and increasingly financialized production systems (Coe 2014). This phase of the development of capitalism relies on the largest metropolitan areas and reinforces their development (Halbert 2021). Neoliberal policies, which aim to extend market mechanisms and disciplines to a growing range of social activities, thanks to State intervention (Pinson and Morel Journel 2016), are being deployed particularly within metropolitan areas (Pinson 2020).

Geographical Places in Transportation,
coordinated by Jean-Baptiste FRÉTIGNY. © ISTE Ltd 2025.

protesters also showed that many of the men mobilized were logistics workers (Collectif d'Enquête sur les Gilets Jaunes 2019).

These events show how much contemporary urban economies rely on the efficiency and deployment of logistics activities. They therefore indicate the significance of these places and jobs for contemporary metropolitan societies. Logistics activities correspond to the range of operations supporting the flow of goods between the different links in the economic system, from places of production to places of consumption. In addition to the problem of flow management, it relates to the concrete, physical operations of transportation, deliveries and warehousing[2].

Indeed, urban dynamics lead to a multiplication of flows of goods in central and peripheral urban spaces. The organization of such flows requires a large number of places dedicated to these activities, mainly warehouses. While often very large within suburban and peri-urban industrial parks, they are sometimes small in size in dense urban spaces, or in historical centers. The accomplishment of these various operations is based on numerous blue-collar jobs, which are today fundamental for the urban working classes (Raimbault 2020a).

Logistics workers are overwhelmingly male, accounting for 90% in 2014[3]. However, subsectors of logistics can feature a high proportion of women, as in the case of the Grand Magasin warehouse, in Marne-la-Vallée (Paris Region), studied by Cuny (2020). Between warehouses, shops, restaurants and consumers' homes, delivery drivers – driving trucks or vans or riding mopeds or bicycles – advance through urban spaces, from the outskirts to city centers. Working directly in the public space (Cholez 2002), they are often subcontractors of large carriers, sometimes digital-platform workers (Vétois and Raimbault 2017).

Blue-collar jobs within the manufacturing industry are on the decline in contemporary metropolitan areas, representing fewer than 15% of blue-collar workers in the Paris Region, for example. But logistics (warehouse and transportation) workers form an increasing share of the blue-collar workers' group, 25% of the total in the Paris Region (Raimbault 2020a). This shift within the contemporary blue-collar workers' group is fully illustrative of a postindustrial turning point that affects workers and, more broadly, contemporary economies and societies. At the same time, evolutions in logistics workspaces contribute to the transformations of contemporary working-class centralities, defined as neighborhoods or municipalities

2 Regarding order picking in warehouses, see Benvegnù and Gaborieau (2017).

3 As such, we choose to refer to these workers using the masculine form.

offering work and revenue, accessible housing and, more broadly, material and relational resources drawn from local anchoring (Rosa Bonheur 2019).

This chapter proposes to decipher the logistics places of metropolitan areas according to a social geography approach. It first analyzes the geography of warehouse workers, which is mainly deployed around the urbanization front of agglomerations. It then looks at the work and places of the delivery drivers and platform workers, mainly articulated around historical centers. In this way, it shows how this geography of postindustrial blue-collar workers and the associated social mobilizations partly reshape the contemporary working-class centralities of 21st-century urban areas.

The analyses are based on the exploitation of population census data (Raimbault 2020a) and surveys, partly collective, conducted around several logistic areas in Seine-et-Marne (Raimbault 2014; Cuny 2020). These surveys were extended with observations on planning policies, conflicts, trade-union strategies and concerning meal delivery drivers, performed notably in Nantes, Vancouver, Luxembourg and Venlo (Netherlands). A review of the literature relating to labor sociology (in particular, Benvegnù and Gaborieau (2017)), geography (De Lara 2018) and studies on mobilities (Aguiléra et al. 2018) completes the exploration work.

4.2. Warehouse blue-collar workers and the social issues of diffuse urbanization

Warehouses are major places of employment for blue-collar workers, whom they are concentrating in increasing numbers. Thus, in France, they numbered 700,000 as of 2014[4]. In comparison, the manufacturing industry employs just under 1.5 million people (Raimbault 2020a). However, the urban-planning logic of logistics parks lead these jobs to be localized in increasingly remote outskirts, breaking with the historical and sociospatial configuration of the manufacturing industry and industrial inner suburbs, which have played a structural role in the history of the working class (Mischi et al. 2013).

The logistics workers' group is therefore confronted with the social and day-to-day difficulties inherent in diffuse urbanization. Very recent social movements indicate a slow, progressive political engagement, or politicization, of these blue-collar employment hubs, which are increasingly emerging as contemporary working-class centralities.

4 Unless otherwise stated, the source for the data in this chapter is the French population census conducted by INSEE in 2014.

4.2.1. *The blue-collar world of warehouses*

Contemporary logistics systems rely on numerous warehouses. Analysis of these places appears essential in order to gain an understanding of the blue-collar world unfolding there. Industrial groups, retail and e-commerce companies, and carriers and couriers (notably responsible for delivering parcels) need sites dedicated to the storage of goods and, above all, the collection and sorting of goods, the packing and unpacking operations and cross-docking (switching from one mode of transportation or type of vehicle to another). While small warehouses still exist, for example, to organize deliveries in city centers, they tend to be larger, up to at least 100,000 or even 200,000 m^2, such as the Amazon, Carrefour and Walmart distribution centers.

The number of jobs assigned to a warehouse depends not only on its size, which is highly variable, but even more on the complexity of the operations to be carried out on goods. While some warehouses represent only a dozen jobs, large e-commerce warehouses can hire several thousand workers.

Since warehouses are often grouped into logistics parks of five or 10 establishments, or even more, the number of jobs generated at the scale of a logistics park is generally in the order of a few hundred, that is, that of an average factory. The largest logistics parks concentrate several thousand jobs, like the large Fordist industrial complexes. In the Paris region, the new city of Sénart, whose economic development is based on logistics activities (Barbier et al. 2019), would bring together nearly 10,000 blue-collar jobs in logistics and transportation of goods in 2014.

In warehouses, workers move, pack, store, sort and label goods. In order to work as a forklift operator, in charge of moving pallets, workers must obtain a specific permit – called the CACES (certificate of operator safety competency) in France. No official qualifications are required for the other positions in the warehouse.

Order pickers produce pallets and parcels, which are then shipped to their recipients. They often work using "voice picking": a software program (directly linked to company-level flow management) directs them, through their headset, as to which goods to take and install on the pallet or to add to the package, and thus their journey through the warehouse. Other workers are stationed along conveyor belts, for example, to fill packages or sort parcels, in the context of increasingly mechanized and automated warehouses.

These different technological devices have led to a very marked intensification of labor[5] (Gaborieau 2016), according to a Taylorist logic of increasing productivity by refocusing work around repeated tasks (Benvegnù et al. 2021). The consequence of this intensification is a soar in these workers' productivity, but also in workplace accidents (Govaere 2009).

4.2.2. *Outer-suburban logistics parks: the emergence of working-class economic centralities beyond the industrial inner suburbs*

Since the 1970s, the development of logistics activities has resulted in the mass construction of warehouses on the outskirts of major metropolitan areas. The geography of these warehouses is relatively similar across most of the world's urban areas (Dablanc and Frémont 2015; Raimbault and Heitz 2024). During a first phase, from the 1970s to the early 1990s, warehouses were mainly built in the industrial and working-class suburbs inherited from the Fordist period, notably within existing industrial parks. In Figure 4.1 relating to the Paris region, these are the municipalities in the "working-class inner suburbs" category. Then, in a second phase, from the 1990s, a new generation of larger and more modern warehouses is built in new logistics areas outside of these old industrial inner suburbs. They are set up in industrial parks developed in the outer suburbs, that is, the most peripheral fringes of agglomerations, or even in the first peri-urban ring. In Figure 4.1, they are municipalities in the category "peripheral employment centers". In these areas, large plots situated close to highways and ring roads are available and are fairly inexpensive.

This logistics sprawl (Dablanc and Browne 2020) marks a real sociospatial rupture from the industrial and working-class inner suburbs, where until then blue-collar jobs were concentrated. The cases of the Paris region (Raimbault and Bahoken 2014), Atlanta (Dablanc and Ross 2012) and the Los Angeles region (De Lara 2018) show that municipalities hosting recent logistics parks are less densely populated than the industrial inner suburbs. Their population is characterized by a somewhat intermediate sociological profile, in which intermediate occupations are overrepresented (Figure 4.1). In addition, right-wing municipalities or conservative majorities are generally predominant, while the historically working-class inner suburbs remain marked by a left-wing vote despite the crumbling of communist and socialist bastions in the case of the Paris region.

5 Intensity further reinforced by the payment of individual productivity bonuses, which are particularly attractive for employees often paid the minimum wage.

Departmental breakdown of the Paris region
HAUTS-DE-SEINE
VAL-D'OISE
YVELINES
PARIS
ESSONNE
SEINE-SAINT-DENIS
SEINE-ET-MARNE
VAL-DE-MARNE
Mantes-la-Jolie
Cergy
Gennevilliers
St-Denis
Aubervilliers
Roissy-en-France
Meaux
Bussy-St-Georges
Châtres
Moissy-Cramayel (Sénart)
Provins
Melun
Nemours
Évry
Rungis
Trappes
Paris
N
0 10 20 km
department border
information not available
Sources: 2012 and 2013 population census: customized cross-tabulation, INSEE [producer], ADISP-CMH [disseminator]
N. Raimbault, S. Charrier

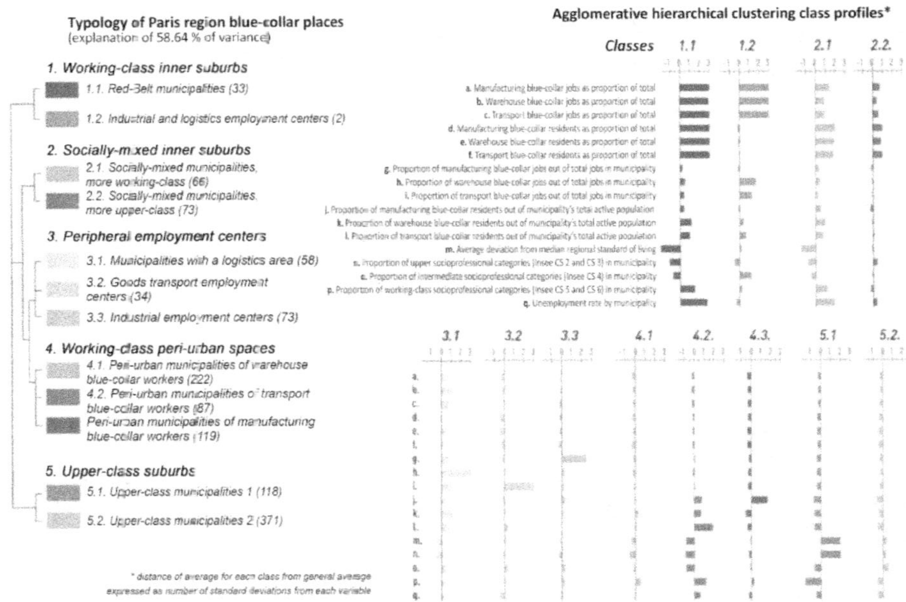

Figure 4.1. *Typology of Ile-de-France blue-collar places (source: Nicolas Raimbault (2020a)). The typology is based on an agglomerative hierarchical clustering (AHC) based on variables concerning the residences and workplaces of the three categories of blue-collar workers (numbers and proportion of the municipality total) as well as the municipality's socioeconomic data (proportions of the upper, intermediate and working-class socioprofessional categories, unemployment rate and average deviation of municipalities from the median regional standard of living). For a color version of this figure, see www.iste.co.uk/fretigny/transportation.zip*

Lastly, the dynamic of logistics sprawl leads to the emergence of new economic centralities for the working classes, in the form of blue-collar employment centers, accessible to people without qualifications, located in the spaces of the urbanization front.

These workplaces structure the career paths of a significant part of the working classes. For these workers, the management of precariousness (temporary and fixed-term contracts) involves anchoring in a logistics park in order to be able to move from one warehouse to another when a contract or assignment comes to an end (Cuny 2020). These centralities are spatially disconnected from the main residential areas of the working classes, which are either in the working-class inner suburbs or in the distant working-class peri-urban spaces, as demonstrated by the example of the Paris region.

The typology of Figure 4.1 differentiates 12 classes grouped into five sets: working-class inner suburbs, mixed inner suburbs, peripheral blue-collar employment centers, working-class peri-urban spaces and upper-class suburbs.

The Paris region blue-collar workers' group is structured around three emblematic areas: the working-class inner suburbs, peripheral blue-collar employment centers and working-class peri-urban residential municipalities. Indeed, the comparison between places of residence and workplaces shows that warehouse workers do not live in the municipalities where these recent logistics parks have been developed (Raimbault 2020a).

The majority still reside in the historically working-class inner suburbs, in the stock of social housing or cheap private housing, for example, in a run-down private residence (Raad 2015). Some are located in the furthest peri-urban ring, which is also the most working-class (Lambert 2015). In both cases, the distance to be covered between these homes and the new logistics parks is large.

In contrast, projects for new logistic buildings, in denser, sometimes mixed, building complexes, are emerging both in the inner and historically industrial inner suburbs, and indeed in the central spaces of agglomerations (Debrie and Heitz 2017).

They can also concern former shops or artisanal premises transformed into small warehouses, known as "dark stores", for "quick commerce" companies and platforms, which offer rapid deliveries of everyday consumer products and meals (Dablanc 2022). This diversity of urban warehouses could also contribute to maintaining blue-collar jobs in the core of contemporary metropolitan areas.

4.2.3. *The social issues of blue-collar labor in diffuse urban areas*

Logistics areas take the form of a juxtaposing of vast buildings with standardized architecture: parallelepipeds covered with metal cladding and corrugated sheet metal. The warehouses are well spaced from each other, surrounded by maneuvering and parking areas for trucks as well as parking lots for staff and visitors. Wide driving lanes run through these areas, punctuated by roundabouts and, increasingly, green spaces added along the lanes and between warehouses. These development principles are eminently functional, dedicated to the handling and movement of goods.

Emblematic of urban planning with functional specialization, the logistics areas offer very few services, whether private or public. These spaces sometimes neighbor retail parks where workers can find several places for dining or socializing. Most often, everyone has to bring their own meal, breaks are taken in the warehouse and everyone leaves as soon as they finish work. Moreover, transit services are more limited in these outskirts. When there is a bus line, schedules are often incompatible with the irregular working hours common to such jobs.

However, for these workers, it is difficult to find low-cost housing near the urban front, where most of the individual housing available is aimed at the middle classes. These difficulties in accessing private housing are exacerbated by the high proportion of precarious contracts in the logistics sector. Consequently, the increasing geographical remoteness of warehouses, from the inner suburbs to the logistics parks on the urbanization front, increasingly accentuates the spatial disconnection between home and workplace for these workers (Figure 4.1). This disconnection makes daily commuting difficult and costly. These findings confirm the analyses of Korsu and Wenglenski (2010) showing, in the case of the Paris region, that blue-collar workers are the socioprofessional group that travels the greatest distance to get to work and is most dependent on automotive mobility due to poor transit connections to their workplaces. However, the costs of using a car are even higher for these workers on low incomes and subject to increasingly precarious conditions (Tranchant 2019). Faced with these difficulties, carpooling is a common mutual-aid solution for these workers (Tranchant 2018). Lastly, the spatial mismatch[6], resulting directly from the logics of logistics park development, makes access to logistics jobs difficult for part of the contemporary urban working classes, and weighs heavily on the daily lives of the blue-collar workers employed there (Cuny 2020).

6 Introduced by Kain (1968) with reference to African American workers in US cities, the notion of spatial mismatch between places of residence and workplaces is presented as an explanatory factor as to difficulties in accessing employment.

4.2.4. *Toward a politicization of these emerging working-class centralities?*

As areas with a high concentration of logistics jobs, logistics parks have, until recently, rarely been understood, and even less appropriated, as working-class centralities, either by the local authorities governing these industrial parks, or by workers themselves.

Due to the disconnection between home and work, the numbers of workers putting down roots are low in these areas, which often lack recognized sites for socializing. Moreover, these logistics parks have very little investment by activist, trade-union, associative and political networks, which remain organized around central cities and historically working-class inner suburbs. Indeed, the unionization rate for workers in the logistics sector is only 4% in France, compared to 10% for all workers combined (Benvegnù et al. 2021). This weak trade-union presence is coupled with fragmentation. From an institutional point of view, warehouse jobs are split between several professional branches: transportation/warehousing for service providers, but also retail and the various branches of manufacturing for warehouses operated directly by large retailers or manufacturing companies. The workers' unions, however, are organized mainly by branches. Lastly, unions are struggling to further establish themselves in warehouses owing to the instability of the workforce, the importance of temporary employment and also the particular geography of logistics sites. FedEx's logistics site in Memphis, for example, is described as "one of the great non-union industrial complexes in the US" (Allen 2015).

Moreover, the main logistics parks are governed by local policies whereby blue-collar workers are absent in every sense of the word. As the workers do not generally live in the municipalities where the logistics areas are located, they are unable to vote in these territories. Beyond this simple observation, the political agendas of these municipalities do not take account of the workers, nor even, often, employment in general. Indeed, the manner in which these logistics parks are produced amounts to the privatization of economic development policies: private real-estate developers-investors choose the companies that set up in the park, not the local governments, as is more or less the case when municipalities themselves develop their own business parks. Thus, the municipalities concerned generally do not have any relationship with the companies in their logistics areas, let alone their employees. Logistics becomes a kind of black box for local governments inside the logistics park: their only relationship to logistics is the real-estate developers-investors. Local political agendas are therefore often reduced to the real-estate issues of interest to these investors (Raimbault 2022). Consequently, logistics parks are rarely thought of as blue-collar employment centers in the local policies of these outer suburbs.

However, several recent dynamics seem to be gradually politicizing these economic centralities for the working classes. Although they were virtually non-existent due to the low unionization rate, as indicated earlier, strikes and conflicts led by logistics workers are on the rise (Cuppini et al. 2015; Benvegnù and Gaborieau 2017).

Figure 4.2. *Picket line at the Leroy Merlin warehouse in Valence (Drôme, France) (source: Khedidja Zerouali (2021), Mediapart). For a color version of this figure, see www.iste.co.uk/fretigny/transportation.zip*

The case of Amazon warehouses is crucial here. In Italy, on March 22, 2021, 9,500 order pickers and 15,000 delivery drivers participated in a 24-h strike across the country (Massimo 2021). The aim was to extend the collective agreement, obtained in 2017 by a warehouse in Piacenza following an initial strike, to all workers in Amazon's logistics chain, including drivers employed by delivery service providers. Since 2013 and the first strike action at a German warehouse, Amazon's various European locations have regularly experienced mobilizations in Germany, France, Italy and Spain. Amazon aside, the Italian logistics sector experienced 14 general strikes between 2013 and 2019, mainly organized by the new so-called "grassroots" unions (Cobas in Italy). They formed within the warehouses of logistics providers in the Po Valley (Milan, Piacenza, Bologna, Modena, Parma, Padua, Turin and Verona), away from the major trade-union confederations (Massimo 2020). Conversely, in 2021, the US Retail, Wholesale and Department Store Union (RWDSU) has failed to establish itself at the vast Amazon warehouse in Bessemer, Alabama, which has nearly 6,000 workers, despite an intense campaign that has resonated internationally. Although successes are still rare and mobilizations vary widely, the logistics sector and the large warehouses of the urbanization front are gradually emerging as key places of contemporary working-class conflicts.

At the same time, warehouses are now commonly identified, during social movements and by activist groups, as economic sites to block in the context of just-in-time flows (Cowen 2014; Alimahomed-Wilson and Ness 2018), as was the case in France during the *gilets jaunes* movement from fall 2018, or during actions against pension reform in late 2019 and early 2020. Indeed, the first surveys and observations, notably those of the Collectif d'Enquête sur les Gilets Jaunes (2019) (collective investigating yellow vests), indicate a significant proportion of logistics workers among yellow-vest protestors. The logistics parks, or the roundabouts that serve them, were therefore well-known sites to the people mobilized during this social movement. They were identified as economic centralities. Their occupation would therefore indicate a progressive political appropriation.

Lastly, these elements highlight that logistics parks are very gradually becoming spaces of social conflict, with mobilizations drawing support from these key capitalism sites of just-in-time flows and a spatial concentration of blue-collar jobs, according to a configuration that recalls the role of factories in the social movements of the Fordist period (Vigna 2007). In this sense, these new blue-collar centralities contrast with the working-class centralities historically constituted within the industrial inner suburbs, which concentrate very varied economic and residential resources and practices, partially peripheral or even autonomous, with respect to post-industrial capitalism (Rosa Bonheur 2019). In the logistics parks of the urban fringes, centrality is structured by production facilities fully integrated into contemporary capitalism (warehouses). These centralities are therefore emerging around labor and economic, social and even political power relations, as evidenced by the social movements mentioned. These sites create visibility with regard to the redeployment of blue-collar jobs to the urbanization front. Sometimes there is a change of scale in mobilizations, as during the strike marking Amazon in Italy in March 2021, which extended to truck drivers. Indeed, the production of logistics flows also involves other workers beyond warehouses and the urbanization front, especially among truck drivers, many delivery drivers who are essential to consumption practices within central areas.

4.3. Delivering the city: working classes and mobile work in urban spaces

Road transportation drivers amount to around 500,000 employees in France (2014), more than half of whom are delivery drivers or couriers, that is, they are in charge of urban deliveries. In addition, this workforce is made up of the self-employed, craftspeople, freelancers and even illegal workers, who are relatively high in number, as we shall see, and who are, by definition, extremely difficult to count. Their working conditions are transformed by the multiplication of deliveries to be performed in urban spaces. Their growing number, difficulties and mobilizations bring visibility to the centrality of manual work in the heart of contemporary metropolitan areas, while contributing to the transformation of working-class neighborhoods.

4.3.1. *An explosion of deliveries, a low-cost economy*

Deliveries are ubiquitous in urban spaces, due in particular to the boom in e-commerce. In France, 20% of the kilometers traveled in cities were due to the transport of goods in 2018. In the Paris region, the growth rate for the transport of goods is twice as high as that for passenger transportation. In the Paris Commune, parcel deliveries account for a quarter of vehicles on the road, and this represent only one part of the transport of goods (Blanquart et al. 2019).

The economy of deliveries, particularly on behalf of e merchants, relies on massive recourse to subcontracting to very small businesses (VSBs) and their employees, or indeed to craftspeople or self-entrepreneurs. In other words, in this link in the logistics chain, salaried workers have historically been bypassed. Delivery drivers are rarely salaried employees of the major courier groups (Chronopost, DHL, FedEx, etc.), who are themselves service providers, on behalf of e-merchants and, more broadly, shipment senders. Under more or less exclusive contracts with courier companies, subcontractors are thus very much under the domination of the contractors.

This amounts in every respect to a low-cost economy. Subcontractors take the commercial risk and have to settle for almost zero margins to the point that many companies are in a permanent state of liquidation, which automatically pulls market prices down. In such a situation, there is substantial recourse to undeclared work (Libeskind 2015).

Delivery routes start, more and more often, within the logistics parks analyzed above, therefore in increasingly remote outskirts. In this sense, these areas also constitute economic centralities for delivery drivers. Smaller warehouses of delivery companies manage nevertheless to remain within the old industrial inner suburbs. The activities of these small warehouses thus also extend the historically working-class centralities. Delivery drivers then travel through agglomerations to deliver to shops and homes, mainly in dense urban spaces. In the course of their work, they encounter a variety of urban spaces, from the outskirts to city centers. If they are to succeed in delivering all their packages, drivers "describe days of more than 12 hours of work, sometimes from 5 a.m. to 8 p.m., six days a week" (Israël 2013). Having to maintain such a high pace means it is generally very difficult for them to comply with traffic rules (parking and speed limits). These very long days of intense urban mobilities also involve handling parcels that are often heavy.

4.3.2. *Delivery platformization*

In parallel to these outsourcing logics, several delivery organization applications have appeared in recent years. Through a digital platform, they assign an independent delivery driver to a trip. The first developments occurred in the field of meal delivery, known as "food tech". Today widely known, these platforms offer a service for ordering meals, among those offered by restaurants registered on the application, and their delivery, by self-employed couriers registered on the application. The food-tech platforms with the largest presence in France are Deliveroo and Uber Eats. Even more recently, according to the same model based on self-employment, several "quick-commerce“ platforms have become established, offering everyday consumer products, delivered to the customer's door within 10 or 20 minutes (Dablanc 2020).

In parallel, platforms offer parcel delivery, connecting shippers with couriers. Certain platforms, such as Stuart (the French *Poste* group), call on self-employed entrepreneurs, in the same way as food-tech applications. Others, such as Colisweb and Deliver.ee, make use of professional carriers, who usually also work as subcontractors to the major courier companies (Blanquart et al. 2019). However, the business volume of these platforms is minimal compared to traditional courier services that continue to rely heavily on subcontractors, according to the logics presented above. Rather than standard parcels, which remain the preserve of couriers, these platforms seem to be further finding their place in the delivery of everyday consumer goods, particularly food. Partnerships are being established between quick-commerce platforms, large retailers and Amazon (Aguiléra et al. 2018).

Initially, a significant number of these deliveries were carried out by bicycle. Unlike the usage of a motor vehicle, which requires a level of professional competence and inclusion on the light carrier register – just as for traditional courier subcontractors – a bicycle delivery can be carried out without requiring any qualification or administrative procedure. However, the reduction in the price of trips, and therefore the increase in production rate, led many delivery drivers to opt for mopeds, but the vast majority of them are used without the required level of professional competence (Dablanc et al. 2020).

Platformization refers to mobilization at the request of self-employed workers through digital platforms. This dynamic does not ultimately fundamentally call into question the current economic logic of deliveries. Rather, it is a digital updating of a system already based on outsourcing (Raimbault 2020b). In the case of salaried work for a subcontractor as in that of self-employed work for a digital platform, delivery drivers are paid per task, according to the "gig economy"[7] principle (De Stefano 2015). They are subjected to economic risk and unsustainable production rates (Vétois and Raimbault 2017). Platformization thus appears in the logical continuation of a substantial trend for outsourcing that began with the liberalization of road transportation. This dynamic seems to mark an ultimate outsourcing of workers in the form of the uberized self-entrepreneur (Abdelnour and Méda 2019). Continuing the logic set in motion by subcontracting, it has substantial day-to-day consequences for delivery drivers.

Platform delivery drivers work mainly within hypercenters, where the majority of restaurants are concentrated and therefore the main places for waiting between orders, which are also places for exchanging information and sociability amongst delivery drivers (Burban 2022). These platforms are also leading to the emergence of new production sites for meals, known as "dark kitchens", and of consumer goods packages, or "dark stores", both in hypercenters and in the former industrial parks of inner suburbs (APUR 2022). In this sense, the platformization of deliveries creates an area for working-class labor in city centers while partially integrated into the historically working-class centralities.

7 Expression from the world of music, where musicians are paid per concert. It corresponds to the situation of self-employed workers paid per task, possibly by multiple contractors, which contrasts with the situation of employees paid monthly by a single employer.

4.3.3. *Self-employed workers and working classes*

Road transportation drivers, and in particular delivery-driver jobs, are classified within the "blue-collar workers'" group according to the INSEE classification of professions and socioprofessional categories (PCS). These positions correspond to a common stage in the career path of working-class men (Cholez 2002)[8]. Beyond salaried employees, the barriers to entry into this market are particularly low for entrepreneurial projects. Many young working-class and immigrant men, especially in the Paris region, try their luck there (Charrin 2013), under the status of self-employed subcontractors or employees of VSBs. In addition, the emergence of digital delivery platforms has multiplied the demand for independent couriers to the point of appearing today as jobs emblematic of the male urban working classes.

Three surveys into meal delivery drivers in eastern Paris were conducted in 2016, 2018 and 2020 (Dablanc et al. 2020). They offer a very valuable portrait of these workers. They show that these activities are now carried out mainly by full-time workers, "for whom it is the only activity and therefore the only source of revenue" (Aguiléra et al. 2018, p. 39). Indeed, full-time delivery drivers went from being a minority in 2016 (23%) to a significant majority in 2020 (73%). For tasks presented at the launch of the platforms as opportunities for "side jobs", students now only represent a small minority (9%). Indeed, the work is intense, with the majority mobilized 6–7 days a week. This is in fact the case for nearly three quarters of full-time delivery drivers, 51% of whom work more than 8 h a day. And yet, delivery drivers are not pursuing a career; turnover is the norm, with length of service rarely exceeding 6 months. For 37% of the drivers surveyed, this is their first job.

The portraits produced by these surveys reveal the great precariousness of a part of the urban working classes. These full-time delivery drivers are almost exclusively male (98% of the total), and 88% are under the age of 35. They hold very few qualifications, rarely beyond high-school diploma level, with 40% having no qualifications, this figure rising to more than half where full-time delivery drivers are concerned. It should be noted that 86% do not have French nationality. Many of them are likely to have recently immigrated to France. Indeed, a significant proportion likely does not have a residence or work permit, as indicated by the large number of shared accounts, itself underestimated due to the significant under-declaration of this type of situation. Account sharing for one digital platform application or another is a common informal practice that allows undocumented people to work illegally for the platforms, generally in return for remuneration paid to the account owner, as

8 As of 2014 in France, 95% of drivers and couriers were men.

confirmed by a survey conducted in Nantes in 2021 and 2022 (Burban 2022). These elements reflect the significance of undocumented labor for a whole section of urban economies. In the Paris region, it is therefore not surprising that delivery drivers live mainly in the working-class neighborhoods of Paris and its inner suburbs. These portraits show that deliveries via platforms "attract people whose youth (and therefore inexperience, lack of networking, etc.), lack of qualifications and/or the discrimination to which they are subjected make the traditional labor market hard to access" (Aguiléra et al. 2018, p. 40). A similar survey conducted in Nantes among 323 delivery drivers in 2021[9] reveals a very similar profile, that of racialized young men working full-time (Burban 2022; Dablanc et al. 2022). These transformations of part of the urban working classes, driven by the emergence of delivery platforms, are therefore also at work in smaller French metropolitan areas than that of the Paris region.

The immediate concerns of delivery drivers relate primarily to the rate for trips, which the platforms, struggling to be profitable (Blanquart et al. 2019), drive downward. They also concern the organization of work by the algorithm, which aims to increase the rate of deliveries, as well as their social security cover, including with respect to workplace accidents: 39% of delivery drivers indicate that they do not have insurance (Dablanc et al. 2020).

These working condition issues are most particularly condensed in the problems encountered on the move, as they pass through places of traffic (Aguiléra et al. 2018). To increase their revenue, delivery drivers multiply deliveries. Their only room for maneuvering consists of being faster, either by violating traffic rules, or by using a motorized vehicle, despite this being prohibited in the absence of a level of professional competence. The "performance" of delivery drivers is enhanced by applications that clearly encourage the multiplication of trips and more broadly require these workers to stay connected and stand ready to work. Under these conditions, 79% of delivery drivers report high levels of road insecurity: 29% have already had an accident (Dablanc et al. 2020).

4.3.4. *Mobilizations and alternatives*

Europe has witnessed a series of mobilizations by platform delivery drivers, like the actions of the Paris autonomous delivery drivers' collective (CLAP: *Collectif des Livreurs Autonomes de Paris*). They are mainly requesting an increase in their

---

9 As part of the Subwork research project, funded by the "Productive City" program of the PUCA (urban development, construction and architecture plan).

revenue, according to a logic of fixed-sum remuneration rather than per trip (Achin et al. 2020a)[10]. While their success may appear very limited, these mobilizations have made these workers visible and highlighted their working conditions in the very hypercenter of the cities concerned. The survey conducted on 323 meal delivery drivers in Nantes indicates that nearly 20% of respondents participated in a strike between 2020 and 2021 (Dablanc et al. 2022).

Moreover, already since the 1990s the multiplication of urban deliveries has raised a number of urban policy issues, from road access and usage to the regulation of the many negative externalities. The situation with moped and bicycle deliveries exacerbates these already existing challenges. In addition to national labor regulations, at local authority level, improving working conditions also involves policies on mobilities and public space management.

The example of a mobilization that began in March 2021 in Nantes is rich in lessons in this respect. Its study is based on work to monitor this movement, still underway at the time of writing of this chapter.

While in 2020 strikes calling for an increase in wages were unsuccessful, at the beginning of 2021, the municipal policy decision to ban motorized two-wheelers in the pedestrian city center was the spark that ignited the situation. The municipal police were instructed to systematically fine those involved. With restaurants being concentrated precisely in this pedestrian area, the motorized ballet of delivery drivers is incessant in the evening, causing significant noise, leading residents to complain. Yet, the organization of deliveries by platforms involves the delivery drivers collecting orders directly from the restaurant entrances. From the first days, a rain of fines fell, reducing the revenue of the delivery drivers.

A mobilization was subsequently organized by the ad hoc "collective of Nantes delivery drivers in conflict" (*Collectif des livreurs nantais en lutte*) (Figure 4.3), supported mainly by trade unions and left-wing political parties[11], addressing its demands both to the municipality and to the platforms. In the short term, the

10 Refer also to the interview with Jerome Pimot, bicycle delivery driver, spokesperson for the Paris couriers' collective, cofounder of CLAP and vice-president of CoopCycle (Achin et al. 2020b).

11 The press release of April 12, 2021, entitled "Solidarity with Nantes two-wheeler delivery drivers. The municipality must hear the demands of the mobilized couriers", has the following signatories: *Union départementale Solidaires* 44, UD CGT 44, SCALA (*Syndicat des coursiers autonomes de Loire-Atlantique*) (CGT), NPA Nantes, *Nantes en commun, Ensemble!* 44, *Féministes révolutionnaires Nantes*, *Lutte ouvrière Loire-Atlantique*, *La France insoumise Loire-Atlantique* and *Union communiste libertaire Nantes*.

collective is calling for negotiations to cancel the fines and arrange set times, as well as parking points to continue the use of mopeds. In the medium term, it asks platforms to provide lower emission vehicles, compatible with the pedestrianization policy. As long as these demands are not heard, delivery drivers are directly and intensely confronted with the contradiction between a demand for almost free and instant home deliveries and the desire for a "safe, clean, sustainable" and policed city (Reigner et al. 2009), both emanating from urban societies.

The situation of platform workers is a borderline case that illustrates, by exacerbating them, the difficulties encountered more widely by all delivery drivers, caught up in this contradiction. The issue of deliveries demonstrates the importance of taking into account the social issues of mobile workers in policies on sustainable urban mobilities.

Figure 4.3. *Press conference of the "Collectif des Livreurs Nantes en Lutte", April 15, 2021 (source: Comité NPA Rosa Luxembourg Nantes (2021))[12]. For a color version of this figure, see www.iste.co.uk/fretigny/transportation.zip*

Lastly, alternatives to delivery platforms are emerging: cooperatives of bicycle couriers have been set up in several European and North American cities. They are mainly grouped within a federation, CoopCycle[13], and form an increasingly integral part of the social and solidarity economy networks. Certain cooperatives are embarking on meal deliveries, in direct competition with platforms. Others propose instead to take on shipments from local, especially craft, companies, mainly using cargo bikes.

12 Screenshot from the video available at: www.youtube.com/watch? v=m5KliFoMcN4 [accessed December 20, 2021].

13 European federation of ecological and socially-responsible bicycle delivery cooperatives: https://coopcycle.org/fr/.

The mobilizations of platform delivery drivers show that social movements are transforming with the changes in the work of urban working classes. These conflicts, which mainly concern immigrant populations, form the link between historical centers and working-class neighborhoods. They thus re-articulate certain political, trade-union and associative resources inherited from the industrial inner suburbs. Faced with the challenges of the platformization of labor and more broadly precariousness, these conflicts ultimately show the importance of places linked to the transportation of goods in the production of working-class centralities in the very core of metropolitan areas.

4.4. Conclusion

The places and flows of logistics are at the heart of the employment and work issues of contemporary (peri)urban working classes. On the one hand, the multiplication of warehouses increasingly concentrates blue-collar jobs in the spaces of the urbanization front, which raises a set of difficulties in accessing work within a context of diffuse urbanization.

Logistics parks, however, are gradually emerging as working-class centralities in the form of employment centers, but also key sites of social conflict. On the other hand, the historical cores of metropolitan areas are passed through by delivery drivers, working often independently, sometimes for digital platforms. In this sense, part of the working classes continue to work in multiple places within these central areas and are increasingly mobilizing there. This dimension of logistics work in city centers is interlinked with the working-class centralities of the inner suburbs, which historically are industrial. For a long time somewhat invisible, the blue-collar worlds of logistics, from deliveries in the hypercenter to warehouses on the urbanization front, are gradually emerging from the shadows, contributing to shaping the working-class centralities of the 21st century.

4.5. References

Abdelnour, S. and Méda, D. (2019). *Les nouveaux travailleurs des applis.* PUF, Paris.

Achin, C., Albenga, V., Benvegnù, C., Delage, P., Gaudillière, J.-P., Le Blanc, N., Roueff, O., Yon, K. (2020a). Éditorial. *Mouvements*, 3(3), 7–10.

Achin, C., Benvegnù, C., Le Blanc, N. (2020b). "L'ubérisation, c'est un projet politique de désyndicalisation", entretien avec Jérôme Pimot. *Mouvements des idées et des luttes* [Online]. Available at: https://mouvements.info/luberisation-cest-un-projet-politique-de-desyndicalisation/ [Accessed 3 November 2022].

Aguiléra, A., Dablanc, L., Rallet, A. (2018). L'envers et l'endroit des plateformes de livraison instantanée. Enquête sur les livreurs micro-entrepreneurs parisiens. *Réseaux*, 6, 23–49.

Alimahomed-Wilson, J. and Ness, I. (2018). *Choke Points*. University of Chicago Press, Chicago.

Allen, J. (2015). Studying logistics. *Jacobin* [Online]. Available at: https://www.jacobinmag.com/2015/02/logistics-industry-organizing-labor/ [Accessed 20 December 2021].

APUR (2022). Drive Piétons, dark kitchen, dark stores. Les nouvelles formes de la distribution alimentaire à Paris [Online]. Available at: https://www.apur.org/sites/default/files/drive_pietons_dark_kitchens_dark_stores_paris.pdf [Accessed 3 November 2022].

Barbier, C., Cuny, C., Raimbault, N. (2019). The production of logistics places in France and Germany: A comparison between Paris, Frankfurt-am-Main and Kassel. *Work Organisation, Labour & Globalisation*, 13(1), 30–46.

Benvegnù, C. and Gaborieau, D. (2017). Produire le flux. *Savoir/Agir*, 1, 66–72.

Benvegnù, C., Gaborieau, D., Rivoal, H., Tranchant, L. (2021). Expositions différenciées et résistances sanitaires : les premiers mois d'épidémie dans les entrepôts logistiques. *Mouvements*, 1, 38–46.

Blanquart, C., Cosperec, A., Vanson-Magalhães Da Silva, E., Zeroual, T. (2019). La rentabilité des entreprises de course urbaine uberisée. *Revue Européenne d'Économie et Management des Services*, 8, 167–186.

Burban, C. (2022). Capitalisme de plateforme et divisions spatiales du travail populaire : géographie sociale des livreurs de repas dans la métropole nantaise. Master's Thesis, Nantes Université, Nantes.

Charrin, E. (2013). *La course ou la ville*. Le Seuil, Paris.

Cholez, C. (2002). Cheminer dans la ville, cheminer dans sa vie. *Espaces et sociétés*, 108/109, 115–130.

Collectif d'Enquête sur les Gilets Jaunes. Bedock, C., Bendali, Z., Bernard de Raymond, A., Beurier, A., Blavier, P., Bonin, L., Courty, A., Della Sudda, M., Delozière, G., Dondeyne, C. et al. (2019). Enquêter in situ par questionnaire sur une mobilisation : une étude sur les gilets jaunes. *Revue française de science politique*, 5(5/6), 869–892.

Comité NPA Rosa Luxembourg Nantes (2021). Conférence de presse des Livreurs Nantais en Lutte [Online].Available at: https://www.youtube.com/watch?v=m5KliFoMcN4.

Cowen, D. (2014). *The Deadly Life of Logistics: Mapping Violence in Global Trade*. University of Minnesota Press, Minneapolis.

Cuny, C. (ed.) (2020). *On n'est pas des robots : ouvrières et ouvriers de la logistique*. Créaphis Editions, Grâne.

Cuppini, N., Frapporti, M., Pirone, M. (2015). Logistics struggles in the Po Valley region: Territorial transformations and processes of antagonistic subjectivation. *South Atlantic Quaterly*, 114(1), 119–134.

Dablanc, L. (2022). La logistique et la ville. Questions environnementales et territoriales de la logistique urbaine. *L'Information géographique*, 86(3), 49–77.

Dablanc, L. and Browne, M. (2020). Introduction to special section on logistics sprawl. *Journal of Transport Geography*, 88. doi: 10.1016/j.jtrangeo.2019.01.010.

Dablanc, L. and Frémont, A. (eds) (2015). *La métropole logistique*. Armand Colin, Paris.

Dablanc, L. and Ross, C. (2012). Atlanta: A mega logistics center in the Piedmont Atlantic Megaregion (PAM). *Journal of Transport Geography*, 24, 432–442.

Dablanc, L., Aguiléra, A., Proulhac, L. Wester, L., Louvet, N. (2020). Enquête sur les auto-entrepreneurs de la livraison instantanée. Research Report, IFSTTAR, Marne-la-Vallée.

Dablanc, L., Proulhac, L., Raimbault, N. (2022). Enquête sur les travailleurs nantais des plateformes de livraison instantanée. Research Report, Projet Subwork (PUCA) [Online]. Available at: https://www.lvmt.fr/wp-content/uploads/2022/03/Livreurs-a-Nantes-des-plateformes-de-livraison-instantanee.pdf [Accessed 3 November 2022].

De Lara, J. (2018). *Inland Shift: Race, Space, and Capital in Southern California*. University of California Press, Berkeley.

De Stefano, V. (2015). The rise of the just-in-time workforce: On-demand work, crowdwork, and labor protection in the gig-economy. *Comparative Labor Law & Policy Journal*, 37, 471–504.

Debrie, J. and Heitz, A. (2017). La question logistique dans l'aménagement de l'Ile-de-France : formulation d'un enjeu métropolitain versus absence de concrétisation dans les projets urbains ? *Géographie, économie, société*, 19(1), 55–73.

Gaborieau, D. (2016). Des usines à colis : trajectoire ouvrière des entrepôts de la grande distribution. PhD Thesis, Université Paris I Panthéon-Sorbonne, Paris.

Govaere, V. (2009). La préparation de commandes en logistique : mutations technologiques et évolutions des risques professionnels. *INRS – Hygiène et sécurité du travail*, 214, 3–14.

Halbert, L. (2021). Ce que les métropoles doivent au capitalisme, et réciproquement. *Regards croisés sur l'économie*, 28(1), 37–45.

Israël, D. (2013). Chronopost : les sous-traitants se disent pris à la gorge. *Médiapart* [Online]. Available at: https://www.mediapart.fr/journal/france/270213/chronopost-les-sous-traitants-pris-la-gorge [Accessed 16 July 2021].

Korsu, E. and Wenglenski, S. (2010). Job accessibility, residential segregation and risk of long-term unemployment in the Paris region. *Urban Studies*, 47(11), 2279–2324.

Lambert, A. (2015). *"Tous propriétaires !". L'envers du décor pavillonnaire*. Le Seuil, Paris.

Libeskind, J. (2015). *La logistique urbaine, les nouveaux modes de consommation et de livraison*. FYP, Paris.

Massimo, F.S. (2020). Des coopératives au syndicalisme de base : la citoyenneté industrielle dans le secteur de la logistique en Italie (1990–2015). *Critique internationale*, 2, 57–78.

Massimo, F.S. (2021). Grève à Amazon en Italie : leçons d'une mobilisation historique. *Contretemps* [Online]. Available at: https://www.contretemps.eu/greve-amazon-italie-travailleurs-syndialisme/.

Mischi, J., Pasquiers, O., Pottier, C., Renahy, N. (2013). Le groupe ouvrier : transformé mais toujours là. *Métropolitiques* [Online]. Available at: http://www.metropolitiques.eu/Le-groupe-ouvrier-transforme-mais.html [Accessed 3 November 2022].

Pinson, G. (2020). *La ville néolibérale*. Presses Universitaires de France, Paris.

Pinson, G. and Morel Journel, C. (2016). The neoliberal city–theory, evidence, debates. *Territory, Politics, Governance*, 4(2), 137–153.

Raad, L. (2015). Les paradoxes de la lutte contre l'habitat privé dégradé à Saint-Denis : favoriser la mixité sociale ou maintenir les classes populaires ? *Métropolitiques* [Online]. Available at: https://metropolitiques.eu/Les-paradoxes-de-la-lutte-contre-l.html.

Raimbault, N. (2020a). Nouveaux emplois ouvriers, nouveaux territoires ouvriers ? Une comparaison des géographies professionnelles et résidentielles des ouvriers de l'industrie et des ouvriers de la logistique en Île-de-France. *Travail et Emploi*, 162, 71–102.

Raimbault, N. (2020b). Le développement logistique des métropoles : des enjeux de régulation exacerbés par le numérique. Working Paper, Sciences Po Cities and Digital Technology Chair [Online]. Available at: https://www.sciencespo.fr/ecole-urbaine/sites/sciencespo.fr.ecole-urbaine/files/2020_06%20-%20Raimbault_compressed.pdf [Accessed 20 December 2021].

Raimbault, N. (2022). Outer-suburban politics and the financialisation of the logistics real estate industry: The emergence of financialised coalitions in the Paris region. *Urban Studies*, 59(7), 1481–1498.

Raimbault, N. and Bahoken, F. (2014). Quelles places pour les activités logistiques dans la métropole parisienne ? *Territoire en mouvement*, 23(24), 53–74.

Raimbault, N., and Heitz, A. (2024). Logistics urbanization, between real estate financialization and the rise of logistics urban planning. In *Globalization and Dynamics of Urban Production*, Aveline-Dubach, N. (ed.), ISTE Ltd, London, and John Wiley and Sons, New York, 73–103.

Reigner, H., Hernandez, F., Brenac, T. (2009). Circuler dans la ville sûre et durable : des politiques publiques contemporaines ambiguës, consensuelles et insoutenables. *Métropoles*, 5 [Online]. Available at: http://metropoles.revues.org/3808 [Accessed 20 December 2021].

Rosa Bonheur (2019). *La ville vue d'en bas. Travail et production de l'espace populaire*. Éditions Amsterdam, Paris.

Tranchant, L. (2018). L'intérim de masse comme vecteur de disqualification professionnelle. Le cas des emplois ouvriers de la logistique. *Travail et emploi*, 155(156), 115–140.

Tranchant, L. (2019). Les ouvriers du déplacement. Travail en entrepôt et recompositions du salariat populaire en France (1980–2018). PhD Thesis, Paris Sciences et Lettres, Paris.

Vigna, X. (2007). *L'insubordination ouvrière dans les années 68. Essai d'une histoire politique des usines*. Presses Universitaires de Rennes, Rennes.

Vétois, P. and Raimbault, N. (2017). L'"uberisation" de la logistique : disruption ou continuité ? Le cas de l'Île-de-France. *Technologie et innovation*, 17(3). doi: 10.21494/ISTE.OP.2017.0142.

Zerouali, K. (2021). Leroy Merlin, des bénéfices record et des salariés en lutte pour 80 euros. *Médiapart*, 28 November [Online]. Available at: https://www.mediapart.fr/journal/economie/281121/leroy-merlin-des-benefices-record-et-des-salaries-en-lutte-pour-80-euros.

5

Public Space and Transportation: Friend or Foe?

Antoine FLEURY[1] and Jean-Baptiste FRÉTIGNY[2]

[1] *Géographie-cités, CNRS, Paris and Aubervilliers, France*

[2] *PLACES, CY Cergy Paris Université, Cergy-Pontoise, France*

5.1. Introduction

Engaging with the field of transportation and its operational concerns has hardly been in the research agendas on public space (Bodnar 2015; Qian 2020). This distancing can be explained by the construction of the category of public space breaking with the technicist and functionalist approaches of the city, long prevalent in transportation and urban development studies. Public space approaches pay less attention to travel and the infrastructure underpinning it than to practices in their plurality, to sociability and to sensory experiences (Figure 5.1).

Public space specialists tend to distance themselves from transportation issues, but they are never far away in their studies. Indeed, public space and transportation are deeply intertwined. Public space corresponds to a network of sites that are potentially – and potentially only – accessible to everyone, giving rise to a myriad of possible interactions and appropriations. As a travel-related assemblage of infrastructures, technologies and practices, transportation is a key condition for accessing and frequenting the sites that constitute public space, while also contributing to their interconnection. However, the relationship between transportation and public space is ambivalent. Though transportation produces public spaces, tailored to its networks (such as highway rest areas and subway stations), it also undermines the existence of other public spaces. The function of

Geographical Places in Transportation,
coordinated by Jean-Baptiste FRÉTIGNY.

transportation can indeed systematically prevail over other uses, as shown by public squares in cities and villages transformed into car parks or bus stations, whose integration into the surrounding public space was not initially envisioned.

This chapter aims to elucidate the tension at stake between these two framings of public space, through its functions, on the one hand, and through its multiple uses, on the other hand. This friction is fueled by the thorny issue of space sharing, or commoning, which poses challenges for designers, users and operators alike. The regulation of this sharing, which is inherently political, is essential to ensuring that transportation takes part of a more comprehensive process of placemaking within and through public space. Rethinking the place of transportation in public space means adopting an alternative approach to transportation, one that focuses on its relations with the city and, more broadly, the social and material world. Transportation history has clearly demonstrated the value of this perspective (Divall and Revill 2009; Baldasseroni 2018).

Figure 5.1. *Acrobatics performance for drivers at a red light in the Roma district of Mexico City (source: Frétigny (2018)). For a color version of this figure, see www.iste.co.uk/fretigny/transportation.zip*

Section 5.2 demonstrates how much transportation and public space are entangled, according to specific modalities. However, these assemblages have widely evolved, particularly since the 1970s. Section 5.3 questions the variability of space sharing between transportation and non-transportation activities. The cocreation of public spaces and transportation has today become a significant challenge, explored in section 5.4. It entails rehabilitating transportation infrastructures that have yet to be recognized as public spaces and achieving objectives of social and environmental justice.

5.2. Transportation infrastructure as an integral part of public space

In its material sense, infrastructure refers to an interdependent set of constructions and developments, including (foot)bridges, stations or roads, understood as spaces dedicated to pedestrian movement and vehicle circulation. Shifting away from a strictly technical understanding of infrastructure requires recognizing it as a staple in the materiality of transportation places, subject to multiple forms of appropriation and embedded within the network of public spaces.

Depending on the scale of the infrastructure, its functional specialization and its material configuration, practiced transportation places involve distinct public spaces that can be grouped into two main categories: point objects that are (im)mobile transportation places, on the one hand, and linear objects that form the road, on the other hand.

5.2.1. *(Im)mobile transportation places as hotspots of social life*

Transportation places as point objects pertain to transit and waiting sites such as stations – whether rail, bus or maritime – as well as airports and highway rest areas. They also include public spaces onboard trains, ferries, buses or shared taxis. They are spaces *par excellence* of copresence and confrontation with otherness, often mythified or decried. They lend themselves to various uses, whether expected or informal, and the public spaces they constitute are a resource routinely mobilized by many actors.

Public space is often created by the transition from one mode of travel to another, or intermodality. It corresponds to an experience of walking and waiting associated with a wide range of potential practices, related to consumption, sociability or strolling, for instance. Istanbul's ferry terminals (*iskeleler*) generate a whole social life, punctuated by the embarking and disembarking of *vapurs* (Figure 5.2). Dense flows of passengers running for their buses or taxis follow on from calmer moments spent waiting. Away from ferry terminals, travelers and local inhabitants cross paths during a walk along the banks or a moment of relaxation in a tea room.

Alongside this visible social life, more surreptitious, norms-bending practices also occur. Prostitution often takes place in station districts, cars or caravans. Flirtation and chatting up, when repressed or played out by minorities, unfold, for example, during car rides for young heterosexuals in Iran (a practice known as "*dor-dor*" in Farsi), and between gay men in highway rest areas for sexual purposes in France and the Netherlands (Gaissad and Deschamps 2007; Mameni 2013).

Figure 5.2. *Urban life around the ferry terminals and on the corniche of Istanbul (source: Fleury (2004)). For a color version of this figure, see www.iste.co.uk/fretigny/transportation.zip*

Some of these transportation places carry a strong symbolic charge. It relates to their historical role in the construction of territories, as shown by major European railway stations. The monumentality of their public spaces clearly indicates their semiotic role as a symbolic place of the city and the country (see Chapter 10): main concourses, glass roofs and canopies, colonnades, etc. When linked to distant horizons, their cosmopolitan nature appears in the multiple tangible signs of heterogeneous populations coexisting. Richard Curtis' widely broadcast film *Love Actually* (2003) attests to these collective representations and practices by opening and ending with a documentary montage of loved ones – from multiple communities – reuniting at the arrival area of long-haul flights at London Heathrow. These images foreground the attachments and emotional intensity at work through these places.

The presence of multiple actors makes of these places resource sites for other protagonists. To live or survive, they engage in practices that can be described as alternative in that they deviate from dominant uses and norms. Street vendors are familiar figures in transportation places, such as in Agadez bus station or in Dirkou (Niger), along trans-Saharan roads, where they converge on vehicles to sell tea, biscuits, candy, cigarettes or other items to passengers (Brachet 2009). These transportation places, notably major stations and subway stations and major airports alike, are also a refuge for homeless people who come there to sleep, beg, socialize or help each other (Damon 1996).

Social life in mobile public spaces contrasts with the sociality of stationary sites, primarily because of the duration and pace set by transportation time itself. Sociability is particularly evident among regulars, as the commuting "seating partners" on the ferry between Newfoundland and nearby Bell Island, who play cards, knit and support each other during storms (Royal and Roseman 2021), or the "train friends" who save seats for each other on the train to Durban and mutually watch everyone's belongings (Preston-Whyte 1982). The encapsulated and constrained nature of the space onboard also animates a game of musical chairs between passengers, shaped by personal preferences and comfort conditions, but also social stratifying forces. In these spaces of imposed copresence, power relationships are indeed fully at play, as in the reproaches directed at Rass, a Senegalese commuter, aboard the train between Florence and Pisa by some travelers when he sits next to them (Frétigny 2011). These public spaces can lead to forms of avoidance and altercations, criss-crossed by racial and gender discrimination or other modalities of oppression.

5.2.2. *Transportation roads and their multiple relationships with public space*

Other public spaces inherently coevolve with transportation as they form thoroughfares. "By virtue of its elementary character and its centuries-old existence, the road can be considered as the archetype of the infrastructure" (Alonzo 2018, p. 17). It becomes a public space as soon as it welcomes a certain diversity of practices, whether through the cohabitation of transportation activities and other road uses, or through intermodality. This burgeoning of public space in the infrastructure can be encouraged by specific planning interventions, be they focusing on sidewalks or the creation of linear parks. Friction in transportation also contributes to the rise of public spaces, as when cars stop at red lights, occasioning a flurry of informal activities, such as windscreen washing, begging or street vending.

The street, as an urban assemblage linking the road to the buildings lining it, is a social, cultural and environmental compromise between traffic and surrounding activities. In particular, a close interdependence is at play between transit and retail, as well as services such as restaurants and cafés. This compromise takes various forms depending on the prominence of traffic. In Ho Chi Minh City's alleys, where 85% of the city's inhabitants reside, traffic is moderate and consists solely of motorbikes. Street vending, children playing, neighbors chatting and cooking on the doorstep are prevalent activities there (Gibert-Flutre 2020). The intensification of traffic in certain alleys participates to their loss of function as public spaces, following a process observed for over 50 years in large cities where car ownership has become widespread.

Conversely, public transportation crystallizes a wide range of social interactions in the street: conflicts regarding the use of spaces, sociability or commercial transactions, near subway stations, bus stops, or when a collective taxi picks up or drops off its customers. More unexpectedly, the possession of individual vehicles (cars and bicycles) fosters professional and social life directly on the pavement through repair services. Informal garages in Parisian working-class neighborhoods offer street mechanics services (Jacquot and Morelle 2019) and, more broadly in the city, app-requested, professional bicycle repairs take place in the open air.

Other traffic sites entail specific links between transportation and public space. Public squares are a case in point, with increased attention to uses beyond transportation, even if they have often been turned into car parks or traffic junctions. Bridges, despite being often engineering structures primarily devoted to traffic, can become promenades with panoramic views, even restaurants, shops or fishing spots, such as the Galata Bridge, spanning the Golden Horn in Istanbul. Compared to traffic sites, these spaces are the subject of significant design work in the fields of architecture and urban planning. Designers, in particular, have imagined new models of landscaped streets from the 19th century (Alonzo 2018). Inspired by European avenues and boulevards, the first North American parkways, designed by Olmsted and Vaux, were created in Brooklyn, around Prospect Park. Lined with trees, these parkways provide wide promenades on either side of the roadway. They serve as traffic arteries, vectors of property value enhancement and connectors between the city's major parks. This landscaped version of the road echoes other development models such as corniches, from Marseille to Beirut and Havana, where city dwellers come to walk, eat ice cream or drink tea while enjoying the view (Assaf 2013).

Figure 5.3. *Varied uses of roads and their surroundings in the urban fringes of Paris Region (French Vexin) (sources: Fleury (2017) and Frétigny (2012)). For a color version of this figure, see www.iste.co.uk/fretigny/transportation.zip*

In the outskirts of urban peripheries, roads are merely an intermittent public space, that of town centers and residential developments, but also agricultural and forest spaces (Figure 5.3) open to walking and cycling (Rougé and Aragau 2019). In French villages of urban fringes, roads are particularly claimed by teenagers at bus stops or around public benches, unlike the village squares, carefully redeveloped but rarely invested by residents. Roads in residential developments, permeated with automobility but with low levels of traffic, also lend themselves to other forms of appropriation, as play areas for children or for car washing. Moreover, this fragmented public space of urban fringes is marked by moments of high intensity, such as garage sales or flea markets. There is thus a high variability over time of the place accorded to activities beyond transportation in public spaces.

5.3. The contested role of transportation in the making of public spaces

Variations in the relative importance given to transportation and other activities, in transportation places and routes, contribute to shaping each public space and the power dynamics among actors. In often constrained sites, the cohabitation of transportation and non-transportation activities gives rise to tensions, including conflicts over space uses, which are part of the life of any public space. Space sharing takes up much of the attention of the designers, operators and regulators of these locales. We will explore these tensions in two ways: first by discussing the levers mobilized to influence the sharing of space between transportation and non-transportation activities, and then by retracing the major shifts in this cohabitation over recent decades.

5.3.1. *Three parameters for sharing space*

Urban design, regulation and operational logics, alongside ideologies and social imaginaries, are three key factors in the distribution of different activities in public space, whether or not they relate to transportation.

First, urban design apportions space by modifying its ground surface (choice of coating, color markings or inscriptions) or by equipping it (barriers and posts, chicanes, seats and benches) (Figure 5.4). For roads, the challenge is to define boundaries for transportation and other uses – for example, with curbs, barriers and bollards on certain roads – or to favor a particular mode of travel – walking in the case of benches, automobile traffic for certain asphalts. Transportation places involve a combination of transportation-specific equipment and a transposition of characteristic amenities of other public spaces: flowerpots, café terraces, kiosks, pianos or food courts. These developments vary greatly, ranging from minimal to sophisticated, durable to temporary.

Figure 5.4. *Shift in space sharing at an intersection in the Cusset district of Lyon (source: Fleury (2018)). For a color version of this figure, see www.iste.co.uk/fretigny/transportation.zip*

The regulation of access and uses also influences this distribution of activities. The organization of public transportation, taxis and the regulation of parking consists of allocating delineated spaces, notably through signage, but also of an arsenal of rules, in particular regarding pricing. Furthermore, a whole set of regulations governs the speed of moving vehicles, sometimes very finely. Speed acts as a major indicator of the public nature of space – understood as its degree of openness and the diversity of its uses. The speed of traffic has considerable effects on the access of the space to certain publics, such as children and people with disabilities, and on its uses, linked to a lesser or greater degree to transportation. More occasionally, legal measures to temporarily close roads to traffic encourage alternative appropriations for the duration of a festivity, a demonstration or a weekend. However, interactions may also involve the application of informal norms, particularly in the Global South, as well as negotiations to allow temporary appropriations (Figure 5.5) or to defuse conflicts, inviting us to be attuned to a broader approach of regulation.

Regulation takes different forms for transportation places, as they are more specialized sites. A varying proportion of space is allocated to trade-related activities (see Chapter 3). Users have less influence regarding the space sharing in these places, as the role of the manager, whether public or private, is more prominent.

Rules are more visible (signage and audio announcements). Technical concerns related to the supervision of passenger and freight flows are particularly pronounced, as are safety and security issues (see Chapter 7). Internal rules and regulations of stations and airports most often prohibit informal sales, political demonstrations and restrict the access to certain spaces to travelers only, limiting idling or accompanying of passengers.

Figure 5.5. *Temporary appropriation of a peripheral street in Old Havana (source: Frétigny (2019)). For a color version of this figure, see www.iste.co.uk/fretigny/transportation.zip*

In addition to this parameter of regulation, the underlying representations, ideologies and imaginations are powerful immaterial settings framing this sharing of activities. These representations contribute to both the inertia of the design of transportation-related public spaces and to tensions when the sharing is questioned. The ideology of the road as a space dedicated to (automobile) traffic is challenged by alternative models drawing on various imaginations of the street and widely shared collective representations. The imagery conveyed by European associations wishing to give the streets back to children conjure up as much the imagination of a village festival as the humanist photography of Robert Doisneau, Sabine Weiss or André Kertész. These alternative ideologies are also sustained by attempts to reorient urban-planning standards as well as technical know-how and professional cultures, reminding us that the sharing of space evolves over time and space according to the different parameters presented here.

5.3.2. *Shifts in the contemporary sharing of public space*

Depending on power relationships, actors attempt to play on different parameters to influence or resist changing the sharing of public space, which we will address here from the 1960–1970s onward. Whether considering roads or transportation places, a dynamic emerges from a dominant functional and technical conception of the space toward a more systematic and explicit consideration of uses, landscape and comfort[1].

5.3.2.1. *"Reclaiming" the street: an obvious transition?*

Since the 1970s, a process toward what is termed the "reclamation" of streets for pedestrians, public transportation and bicycles has been observed simultaneously in multiple places in Western Europe (see Chapter 6). A recent manifestation of this change is the resurgence of trams in Swiss and French cities. Since the 2000s, they have been heavily used to diversify transportation modes and redevelop public spaces – including through artistic interventions (sculptures, murals or sounds to announce stations). This paradigmatic example should not obscure more diffuse, and perhaps deeper, transformations that have changed the cities of North-West Europe, following the "residential courtyard" (*woonerf*) model created by Dutch designers. This model integrates children's play areas and neighborhood sociability into a shared space that also accomodates car traffic. Vehicle speed is reduced and through traffic is severely limited. This roadway sharing is accompanied by a wider redesign of public space in residential neighborhoods: changes in pavement, urban greening, addition of benches, playgrounds or fountains. Implemented in the Netherlands from the 1960s, it quickly spread to many Danish, German, Swiss and Austrian cities. It was then adapted in the 1990s–2000s as "home zones" in London and other British cities (Olagnier 2006), and later in France as "meeting zones" (*zones de rencontre*), under the impetus of public bodies such as the Commission for Architecture and the Built Environment (CABE) in the United Kingdom and the Center for Studies and Expertise on Risks, the Environment, Mobility, and Planning (CEREMA, formerly CERTU) in France.

In Southern European cities, this "reclaiming" is deeply linked to an urban culture and imagination highly focused on public spaces, squares especially. In Bologna, as early as the late 1960s, the establishment of limited-traffic zones in central areas was coupled with efforts to restore and highlight the built and landscape heritage. Pedestrians, cyclists, as well as students and tourists, concomitantly reinvested public spaces. The resulting revival of liveliness –

1 Regarding mobile transportation places, which have been the subject of few studies, the suitability of this dynamic is yet to be confirmed.

including restaurant terraces, festive events, etc. – foreshadowed an evolution that many Italian and other European cities would experience in the following decades (Gomes 2017). Another pioneering city in this movement is Barcelona. At the end of the Franco era, elected officials engaged the city in a democratic redesign of public spaces, involving citizens in decision-making. They paid attention not only to the famous Ramblas but also to residential neighborhoods located outside the city center and to their squares. This transformation, which became a staple in urban planning, was notably transposed to Lyon in the 1990s.

These various trajectories show that these dynamics occur over long periods of time, linked as much to local sociopolitical contexts as to the mobility of planning models. They are often undertaken by elected officials who aim to mark a political and urban break upon their accession to power. The gradual mutating of professional cultures and designers' imaginations is also a key factor of change, as is the slow diversification of actors involved in the production of roads. It involves a reorganization of the internal functioning of administrations to overcome the technical and administrative compartmentalization of different departments (roads, parks and recreation, cleaning, urban planning, etc.), opening up to architects, landscapers and even social science experts, relativizing the predominant role of engineers (Fleury 2007; Gomes 2017). But the role of civil society in driving these transformations is equally crucial, whether through opposition to major road infrastructure projects in the 1970s or more recent support for cycling and pedestrian facilities promoted by user associations. Conversely, other fractions of civil society defend automobile access to the city, such as associations of motorists and retailers, opposed to these latest developments, as when the Parisian riverbanks and the center of Brussels were pedestrianized.

These well-identified trajectories are observable in many ways in different cities around the world, particularly in cities on the East and West coasts of the United States and Canada, though it would be misleading to consider them as a universal phenomenon. Developments in the Global South, as understood through the relatively limited number of available studies, invite even further to challenge the obviousness of this "reclaiming". Indian cities stand in contrast, for example, due to the minimal public regulation of street uses and the high social prestige associated with car ownership. The 1990s marked the advent of public-private partnerships, favoring cars through the construction of road viaducts and highways, or exclusive mass transit systems, particularly subways, for the wealthier classes. These projects have encroached on the mobility of pedestrians, despite them undertaking the majority of trips, as well as of cyclists and rickshaws, and have contributed to the marginalization of street vendors. In the absence of formal regulation, trucks and cars dominate the streets, without completely barring other uses (Jain and Moraglio 2014). These trajectories warrant further study and deeper understanding.

5.3.2.2. *Transportation places: toward comfort under control?*

The evolutions of transportation places are also marked by a shift from a functional logic to one that emphasizes the user experience and travel comfort. This is particularly noticeable in the case of the Paris subway since the 1980s. The in-house expertise in social sciences developed within its operator (RATP) contributes to inflecting operational representations on the subway space, from a separate space to one integrated into the city. This shift reflects a new perspective on the network space, recognizing its quality of public space. Following a certain vision of improving passenger comfort, the operator transforms its interventions in the design, management and regulation of the space (Froment-Meurice and Fleury 2016). The renovation of stations considers the sensory dimensions of travel (auditory, visual and olfactory). RATP has introduced a policy of cultural and commercial animation, notably by selecting musicians who are authorized to perform on the subway (Figure 5.6). Inspired by the Toronto and New York commissions established in 1980 and 1985 respectively, the licensing body (Espace Métro Accords) was created in 1997. These measures, combining concerns for traveler comfort with the operator's desire to control public space, have led to an increasing crackdown on individuals deemed "undesirable" (homeless people, beggars, street vendors, unlicensed musicians, etc.), who are perceived as factors of insecurity.

The attention paid to comfort is accompanied by a diversification of activities that go beyond transportation, relativizing the functional specialization of these places. The growth of shops and services illustrates this well, especially with regard to train stations. In Japan, these places have certainly long been designed to be multifunctional, notably in suburban areas, where they play a key role of centrality (Aveline 2003). However, the expansion of their commercial functions has intensified since the 1950s with the emergence of shopping malls, and further accelerated from the mid-1970s as the commercial and office centrality of station districts strengthened. This model has been widely disseminated in Europe since the 2000s in the context of railway station renovation and extension projects. These projects have led to the disappearance of concourses, in favor of more convivial spaces, but with predefined uses, often linked to shopping and food. The spaces and seats associated are subject to greater surveillance, contributing to increasing the normativity of public spaces.

Figure 5.6. *Licensed musician and stage installed by RATP in the Paris subway (source: Froment-Meurice (2016). For a color version of this figure, see www.iste.co.uk/fretigny/transportation.zip*

This surveillance is particularly evident in airports. These public spaces became open to a large public in the 1960s, illustrated by Gilbert Bécaud's song *Dimanche à Orly* (Sunday at Orly) and the crowded terraces of the Parisian airport, France's most visited monument in 1965. Airports have since been marked by a major expansion of spaces dedicated to shops, but also by a dynamic towards closure from the 1970s, as part of anti-terrorist, migration and health policies (Pascoe 2001). They have become laboratories for experimenting with surveillance techniques later applied in other public spaces: video surveillance, security gates, the collection and centralization of personal data, etc. The partitioning at work, between passengers and non-passengers, systematic in airports, is increasingly observed in major railway stations. This dynamic of closing-off is also taking place on a larger scale with the trend toward militarizing airport fences and ferry terminals, contributing to their growing detachment from the city (Cosgrove 1999; Guenebeaud 2017).

This comfort under control that characterizes transportation places is found in many ways today in the streets and squares of numerous cities around the world (Mitchell and Staeheli 2005; Morange et al. 2018). As for transportation places, the "reclaiming" of roadways falls within a neoliberal logic of attracting investors and the upper classes while also contributing to the eviction of minorities. The ambiguity of these developments raises the question of a fair co-construction of public spaces and transportation.

5.4. Designing public space and transportation together

The linkage between public spaces and transportation today raises many questions in its implementation. The first challenge lies in the tangible development of a fair balance for the different actors, human and non-human, and their uses, especially at a micro-scale. A second challenge for our societies is the extension of the public space domain to transportation infrastructures that are still perceived in technical terms, forming a new frontier in recognizing the publicness of these spaces.

5.4.1. *The quest for a fair sharing of space*

Whatever the place given to transportation, roads or transportation places are, in their development, part of the creation of a public space involving multiple protagonists and mutual adjustments. A good example stems from the design of public spaces around the stations of the Grand Paris express network, a project of six new or extended subway lines in the city region. They are qualified as "Grand Paris squares", and acting as intermodal interchange hubs, they bring together multiple actors. The institution responsible for the construction of the network and stations (Société des Grands Projets) and the transportation authority (Île-de-France Mobilités), more focused on intermodality, have divergent approaches. Yet, they have agreed on the importance of "making room", that is, creating as vast an open space as possible, even though every square meter in the area is highly sought after. This perspective aims to promote all possible intermodalities (buses, bicycles, walking, etc.) and a wide range of uses associated with intermodality and the comfort of (non)travelers. A design guide details these principles (TVK et al. 2019) aimed at future contracting authorities (local authorities), who are the ultimate decision-makers. It encourages them to make these squares more hospitable by focusing on waiting, resting or playing, providing ample and diverse seating, as well as water points and public toilets, while also incorporating permeable surfaces and vegetation whenever possible. At the local level, these principles are sometimes the subject of lengthy discussions, especially where local authorities defend a status quo on space sharing and the place of the automobile. Similar ideas have nevertheless inspired the design of public spaces adjacent to the main train stations of Haarlem in the Netherlands or Nantes in France (Figure 5.7).

The question of space sharing also arises in the design of transportation places, which involves delicate negotiations among stakeholders. In the case of airports, the width of the corridor within the retail area, which passengers must navigate to reach their boarding gate, is the subject of intense discussions between commercial operators, on the one hand, and those responsible for airlines' commercial operation, as well as safety and fire prevention officials, on the other hand (Frétigny 2013). The former favor a narrow pathway with a curvilinear layout to encourage passengers to slow down, see more products and consume. In contrast, the latter seek smooth access to the plane and, in the event of an emergency, to escape routes.

Figure 5.7. *The new northern forecourt of Nantes central station, inaugurated in 2019: an open-air waiting room (source: Fleury (2021)). For a color version of this figure, see www.iste.co.uk/fretigny/transportation.zip*

Power relationships and social justice issues fully participate in this space sharing. Policies in favor of cyclists and pedestrians in urban centers can prove ambiguous. The "My street" urban-development operation in Moscow, for example, officially aimed to promote a comfortable city, but primarily reinforced existing corruption dynamics in the public works sector, diverting substantial funding from the rest of the country and its development (Trubina 2020). The poorly laid granite pavers of the sidewalks were very costly, but also require yearly repairs benefitting contractors and related actors. Paradoxically, the installation of this new surface impeded walking on the sidewalks due to the disruption caused by the repair work and the slippery nature of the pavers.

With regard to cycling policies, the implementation of bike-sharing systems in public spaces often imply public-private partnerships. They are contracted with large transnational firms that also manage other urban furniture (advertising displays in the case of JCDecaux), participating in a wider encroaching on public space. These systems also allow these companies to gather data on the uses of urban space, which they can subsequently monetize as unique expertise (Huré 2022), highlighting the growing role of digital technology in the relationship between public space and transportation.

Beyond the allocation of space among social uses, the question of the space devoted to non-human life, particularly plants, in transportation-related public spaces is increasingly pressing in the face of the climate emergency and the biodiversity crisis. Transportation places are increasingly subject to operations that are sometimes spectacular but often unecological in response to the expectations of travelers or corporate social and environmental responsibility mandates: installation of plants and waterfalls, such as the indoor Jewel garden at Singapore-Changi airport, or the use of wood as a visible construction material in terminals. More crucially, more and more streets and squares are being transformed to give more space to non-human life. While already partly integrated into professional cultures, non-human life calls for new expertise, with contributions from naturalists and landscapers in particular reshaping the design of the road. Eco-neighborhoods act as a showcase and laboratory for these transformations (Valegeas 2014). In these neigborhoods, the space allocated to cars is limited, for example, by lining streets with grassy ditches (swales), absorbing runoff; by providing few parking spaces; and by introducing new forms of public spaces such as alleyways and tree-lined avenues.

These layouts slow down automobile speed, promote active mobilities and the diversity of uses, and make public spaces more permeable, vegetated and aquatic. The challenge is to take into account the associated ecosystem services, in terms of limiting climate change by encouraging modal shifts, improving walking comfort, but also enhancing biodiversity, as well as limiting urban heat islands, preventing floods risk, purifying wastewater, decontaminating soils and reducing traffic noise. However, the gap between these public spaces and certain uses – such as parking in surrounding neighborhoods – causes tensions with residents of these neighborhoods, as in the case of the GWL Terrein eco-neighborhood in Amsterdam (Boisonnade 2011).

Many public spaces tend to be gradually transformed at a micro-scale according to this model, whether or not they are part of ecological continuities. Asphalt-removal operations correspond to a diffuse transformation targeting fractions of the road withdrawn from parking or automobile traffic in favor of vegetation. In Montreuil, next to Paris, the municipality has carried out a program in this vein called "Small alternative public spaces" (PEPA) involving local residents (Figure 5.8). These policies are however faced with multiple technical constraints, particularly those related to underground infrastructure (underground car parks, subway stations or utility networks). Moreover, the installations complicate the management of the spaces to the point of becoming a major issue for these actions. The costs of maintaining living elements prove significant, particularly tree pruning in streets and car parks. Local government departments must coordinate more closely, and certain local residents voice their incomprehension at this new model of public space, sometimes perceived as unaesthetic or dysfunctional.

Figure 5.8. *Street garden maintained by residents in Montreuil (eastern suburbs of Paris) (source: Fleury (2018)). For a color version of this figure, see www.iste.co.uk/fretigny/transportation.zip*

5.4.2. *Extending the domain of public spaces to other transportation infrastructures*

Even specialized transportation spaces, infrastructures generally regarded as lacking distinguishing features and with little access to pedestrians, are increasingly recognized as public spaces. This process began with efforts to integrate new infrastructures into the urban fabric, but also by rehabilitating the sidelines of transportation infrastructures and their uses. By the late 1990s, designers such as Mangin and Panerai (1999, p. 70) considered that "public transport or urban highways [...] lead to unbuildable zones than can be more than just neglected areas". They take as an example the paseo de Colón in Barcelona. This road combines a heavy-traffic underground passage and an open-air expanse of public space connecting the city to the port, including a bus-only lane.

This approach is reminiscent of the ring-road covers such as those undertaken for the Paris *périphérique*. It does not call into question the place accorded to traffic, nor the nuisance generated, whether noise or air pollution, which persist on a larger scale. Beyond the waterfront redevelopments during the 1980s–1990s in the field of port infrastructure, more radical and spectacular transformations of infrastructures are emerging, with the repurposing of old railway rights-of-way, such as New York's High Line, the planted promenade and the small belt of Paris, but also of highways. Seoul is a reference, by dismantling in 2005 a viaduct covering the Cheonggyecheon River, and transforming another, Seoullo 7017, into a sky garden in 2017. These changes however are frequently envisaged in a more progressive and forward-looking manner, while maintaining transportation functions. The 2014 master plan for the E40 highway in Brussels outlines a strategy for reinvesting and gradually "recycling" the infrastructure, transforming it into a parkway and prioritizing a broader range of uses and mobilities (TVK 2015) (Figure 5.9).

The transformation also builds on experiments in tactical urbanism, which entails micro-scale actions, engaging users, using minimal resources, and testing them in the short term to achieve long-term, deeper transformations. In New York City, the city's Department of Transportation transformed certain crossings on Manhattan's main avenues (starting with Times Square in 2009) as part of its Plaza Program (Figure 5.10). The use of paint and street furniture (flowerpots, seats and tables) aims to pedestrianize part of the roadway and promote new uses, with the intention of later making more permanent changes (Sadik-Kahn 2016). The open-ended nature of such projects is particularly evident in the case of Berlin's Tempelhof, the former West Berlin airport closed in 2008 (Figure 5.11). Opened to the public as a temporary park in 2010, the unbuilt rights-of-way of the airport were to be transformed into a residential and commercial district, but a citizen mobilization led the Berlin authorities (the *Land*) to halt the project and turn it into a space of "careful" (*behutsam*) co-construction.

Figure 5.9. *Project to gradually transform the E40 highway in Brussels into a parkway (source: TVK (2015)). For a color version of this figure, see www.iste.co.uk/fretigny/transportation.zip*

Figure 5.10. *Madison Square in Manhattan, after its redevelopment as part of the NYC Plaza Program (source: Fleury (2015)). For a color version of this figure, see www.iste.co.uk/fretigny/transportation.zip*

Figure 5.11. *Berlin Tempelhof airport and its "careful" redevelopment project (source: Fleury (2010)). For a color version of this figure, see www.iste.co.uk/fretigny/transportation.zip*

Many spaces are still largely unrecognized as public spaces. Car parks are primarily designed from the perspective of car travel, overlooking other uses and mobilities, especially pedestrian. However, the intermodality at play gives them significant potential for publicness. In residential areas, the car parks are used for a whole range of activities, as shown by Lefrançois (2014) for large public housing units around Paris: mechanics, DIY, lounging, sociability among young people, etc., therefore attracting a wide range of residents. Shopping mall car parks are also subject to alternative appropriations, with young people playing games, without consuming, as in Ho Chi Minh City (Gibert-Flutre 2019). They are also occupied by informal, budget-friendly cafés, where saleswoman and certain customers from the shopping malls gather for lunch, rather than at Starbucks. These often-invisible uses contrast with activist interventions of temporarily repurposing parking spaces in urban centers, transforming them into green and recreational areas. The San Francisco REBAR collective, composed of designers, creators and activists, initiated the international Park(ing) Days movement in 2005, which has since become an annual event attended by a wide network of local activists (Douay and Prévot 2014). Inspired by mobilizations in Bogota, the citizens' initiative Open Streets Cape Town has organized regular street closures to cars in various neighborhoods since 2012, aiming at creating shared spaces "that can help bridge social and spatial divides" inherited from apartheid (Cape Town Travel 2018). This type of initiative was subsequently institutionalized in many cities. Working with local residents (including shopkeepers, school heads, etc.), public authorities are gradually transforming streets and squares to provide more places to eat, drink and discuss with others. Inspired by REBAR's actions, "placottoirs" (a portmanteau formed from "*placoter*" – to chat in Quebec French – and "*trottoir*", meaning sidewalk), that is, "places to chat", thus extend the surface of the sidewalk with a platform that includes plants, benches, tables or bike racks. They have mushroomed in Montreal since 2014 under the impetus of certain boroughs, issuing seasonal occupancy permits to shopkeepers, restaurant and café managers, provided they comply with certain rules.

Industrial or business parks are also pioneering sites where the recognition of the publicness of transportation-related spaces is at stake. While these are daily workplaces for many employees, their design is limited to facilitating car traffic and parking for both personal vehicles and company trucks and vans. Despite the absence of sidewalks and the inhospitality of such places (lack of street furniture or suitable lighting, etc.), employees use them for lunch or a break. Tentative strategies to reorient this development model emerge, such as in Paris-Charles-de-Gaulle airport area, where the airport operator, Aéroports de Paris, is considering diversifying and densifying activities to encourage alternative mobilities to cars at a micro-scale, and the development of lanes reserved for active mobilities. However, the reluctance of companies and the scale of the transformations needed to bring about change tend to slow efforts aimed at freeing these spaces from the functional model attached to them. If these spaces are to become public spaces, they need to be reconnected on a larger scale to other public spaces by non-automobile mobilities. The challenge is to build continuous public spaces by (re)opening paths and redesigning roads (sidewalks, bike paths and lighting). The Covid-19 health crisis has accelerated local reinvestment in existing trails, as in England, where the creation of a comprehensive map is used by activists as a tool to advocate for changes to existing regulations, which greatly restrict access to agricultural and forest areas (Slow Ways 2021). However, these developments are hampered by major land-use issues, tied to the high value of land in urban fringes and to the diffidence of land owners (Banos and Sabatier 2010).

5.5. Conclusion

This chapter highlights the full value of taking a comprehensive approach to all transportation-related public spaces, whether roads or (im)mobile transportation places. Indeed, these spaces are highly interdependent. Beyond differences in land ownership status, the practices of individuals engage continuities from one public space to another, from the train station to the station district, or from the shopping center to its surroundings (car parks, services and nearby residential areas, etc.). Urban-planning and management models also circulate. The flow management and control developed for airports and subways have been widely applied to other transportation-related public spaces, such as street furniture designed to deter the prolonged presence of homeless people, initially tested in the subway, or the introduction of queue systems, from departure lounges to museum forecourts, on sidewalks adjacent to cinemas, or even schools and shops, in the context of the Covid-19 health crisis.

Regardless of the public space, the respective place of transportation and other uses is the subject of a social, political and environmental compromise over the short or long term. Its evolutions are often slow, due to the gradual changes in professional cultures, institutional functioning and individual routines, as well as the material configuration of the space, which can prove difficult to transform. A whole part of activities in our societies revolves around negotiating space between transportation, other social activities and the presence of living beings, a process to which architects and urban planners such as Cerdà, Alphand, Olmsted and Vaux have greatly contributed. The various stakeholders demonstrate inventiveness in constructing context-specific spatial arrangements, shaped by social expectations, the techniques of their time as well as the practices and imagination of places.

Addressing ecological issues will undoubtedly constitute a major challenge in the coming decades for spaces where competition for use is high.

These compromises involve major power dynamics, which gives them an ideological dimension. They raise the question of the role taken up by the different actors, starting with citizens, in the definition of space sharing. The notion of "reclaiming" encapsulates the ambivalence of any transformation: improving comfort while excluding the "undesirable"; creating more space for pedestrians and cyclists but within a strategy aimed at attracting solvent residents, tourists and high-end service companies; promoting shared mobility to reduce reliance on private cars, yet increasing the influence of private operators on public spaces, etc.

These "reclaiming" initiatives form an active part of social changes to neighborhoods, in particular through what has been termed "green gentrification" (Argüelles et al. 2021; for broader discussion on these issues, see also Reigner et al. (2013)). Challenging the dominance of cars and technocratic approaches to transportation places, they concern certain urban centers, while most urban and rural spaces tend to escape these evolutions (Valderrama Pineda and Vogel 2014). Taking transportation places, streets and other roads seriously as key resources and meaningful sites for many citizens represents a major social and political issue. It urges us to look beyond the hypercenters of major cities in the Global North, and to ensure real participation from each and every one in the management and planning of all spaces, including the less visible ones.

5.6. References

Alonzo, E. (2018). *L'architecture de la voie. Histoire et théories*. Éditions Parenthèses, Marseille.

Argüelles, L., Cole, H.V.S., Anguelovski, I. (2021). Rail-to-park transformations in 21st century modern cities: Green gentrification on track. *Environment and Planning E: Nature and Space*. doi: 10.1177/25148486211010064.

Assaf, L. (2013). La Corniche d'Abu Dhabi : espace public et intimités à ciel ouvert. *Arabian Humanities*, 2 [Online]. Available at: https://journals.openedition.org/cy/2625 [Accessed 9 July 2021].

Aveline, N. (2003). *La ville et le rail au Japon. L'expansion des groupes ferroviaires privés à Tôkyô et Ôsaka*. CNRS Éditions, Paris.

Baldasseroni, L. (2018). La piétonnisation des rues, modèle pour des espaces urbains standardisés ? Réflexions lyonnaises, années 1970-1980. *Annales de la recherche urbaine*, 113(1), 12–23.

Banos, V. and Sabatier, B. (2010). Les espaces périurbains non bâtis en France : entre publicisation "urbaine" et privatisation "rurale" ? *Articulo - Journal of Urban Research*, 3 [Online]. Available at: http://journals.openedition.org/articulo/1524 [Accessed 9 July 2021].

Bodnar, J. (2015). Reclaiming public space. *Urban Studies*, 52(12), 2090–2104.

Boissonade, J. (2011). Le développement durable face à ses épreuves. Les enjeux pragmatiques des écoquartiers. *Espaces et sociétés*, 147(4), 57–75.

Brachet, J. (2009). *Migrations transsahariennes. Vers un désert cosmopolite et morcelé (Niger)*. Éditions du Croquant, Bellecombe-en-Bauges.

Cape Town Travel (2018). Open street cape town [Online]. Available at: https://www.capetown.travel/open-streets-cape-town/ [Accessed 22 September 2022].

Cosgrove, D.E. (1999). Airport/landscape. In *Recovering Landscape: Essays in Contemporary Landscape Architecture*, Corner, J. (ed.). Princeton Architectural Press, New York.

Damon, J. (1996). La gare des sans-abris. Un miroir de la question sociale. *Annales de la recherche urbaine*, 71, 120–126.

Divall, C. and Revill, G. (2009). Les cultures du transport : représentation, pratique et technologie. In *De l'histoire des transports à l'histoire de la mobilité ?*, Guigueno, V. and Flonneau, M. (eds). Presses Universitaires de Rennes, Rennes.

Douay, N. and Prevot, M. (2014). Park(ing) day : label international d'un activisme édulcoré ? *Environnement urbain/Urban Environment*, 8, 14–33.

Fleury, A. (2007). Les espaces publics dans les politiques métropolitaines. Réflexions au croisement de trois expériences : de Paris aux quartiers centraux de Berlin et Istanbul. PhD Thesis, Université Paris 1 Panthéon-Sorbonne, Paris.

Frétigny, J.-B. (2011). Habiter la mobilité : le train comme terrain de réflexion. *L'information géographique*, 75(4), 110–124.

Frétigny, J.-B. (2013). Les mobilités à l'épreuve des aéroports : des espaces publics aux territorialités en réseau. PhD Thesis, Université de Paris 1 Panthéon-Sorbonne, Paris.

Froment-Meurice, M. and Fleury, A. (2016). Orchestrer la présence des musiciens dans le métro parisien. *Géographie et cultures*, 98, 113–134.

Gaissad, L. and Deschamps, C. (2007). Des sexualités dans l'espace public. *Espace populations sociétés*, 2(3), 357–369.

Gibert-Flutre, M. (2019). *Les envers de la métropolisation. Les ruelles de Hồ Chí Minh Ville, Vietnam*. CNRS Éditions, Paris.

Gomes, P. (2017). La production de l'espace public à Lyon, Lisbonne et Louvain-la-Neuve : politiques, processus et prise en compte des usages. PhD Thesis, Université Paris-Est, Créteil.

Guenebeaud, C. (2017). Dans la frontière. Migrants et luttes des places dans la ville de Calais. PhD Thesis, Université de Lille 1, Lille.

Huré, M. (2022). Mobilités partagées et régulations politiques. Essai sur les mutations de l'espace public. In *Les espaces publics à l'épreuve des mobilités*, Fleury, A., Frétigny, J.-B., Kanellopoulou, D. (eds). Presses Universitaires de Rennes, Rennes.

Jacquot, S. and Morelle, M. (2020). Mécanique de rue en banlieue parisienne : centralité populaire et migrations. In *Le petit commerce dans la ville-monde*, Fleury, A., Delage, M., Endelstein, L., Dubucs, H., Weber, S. (eds). L'œil d'or, Paris.

Jain, A. and Moraglio, M. (2014). Struggling for the use of urban streets: Preliminary (historical) comparison between European and Indian cities. *International Journal of the Commons*, 8(2), 513–530.

Lefrançois, D. (2014). *Le parking dans les grands ensembles*. Éditions de la Villette, Paris.

Mameni, S. (2013). Car flirting and morality cruising: Neurotic gazes and paranoid glances in contemporary Iranian art. *Al-raida*, 141/142, 27–38.

Mangin, D. and Panerai, P. (1999). *Projet urbain*. Éditions Parenthèses, Marseille.

Mitchell, D. and Staeheli, L.A. (2005). Clean and safe? Predevelopment, public space, and homelessness in downtown San Diego. In *The Politics of Public Space*, Low, S. and Smith, N. (eds). Routledge, New York.

Morange, M., Pilo', F., Spire, A. (2018). Experiencing regularisation in Accra, Cape Town and Rio de Janeiro. The actual right to the city, neoliberalisation and everyday life. *City*, 22(5/6), 685–702.

Olagnier, P.-J. (2006). Les "nouvelles" pratiques d'aménagement des rues des villes britanniques. Les Home Zones : la variante anglaise des expériences de voirie partagée. *Flux*, 66/67(4), 39–49.

Pascoe, D. (2001). *Airspaces*. Reaktion Books, London.

Preston-Whyte, E. (1982). Segregation and interpersonal relationships: A case study of domestic service in Durban. In *Living Under Apartheid*, Smith D.M. (ed.). George Allen & Unwin, London.

Qian, J. (2020). Geographies of public space: Variegated publicness, variegated epistemologies. *Progress in Human Geography*, 44(1), 77–98.

Reigner, H., Brenac, T., Hernandez, F. (2013). *Nouvelles idéologies urbaines. Dictionnaire critique de la ville mobile, verte et sûre*. Presses Universitaires de Rennes, Rennes.

Rougé, L. and Aragau, C. (2019). Appropriations, partages et fabrications de l'espace public. Vers un périurbain plus convivialiste ? *Revue du MAUSS*, 54(2), 307–319.

Royal, D. and Roseman, S.R. (2021). Co-passengering and the gendering of a mobile ferry space. *Journal of Transport Geography*, 92. doi: 10.1016/j.jtrangeo.2021.103000.

Sadik-Khan, J. and Solomonow, S. (2016). *Streetfight: Handbook for an Urban Revolution*. Viking, New York.

Slow Ways (2021). Help create a national walking network [Online]. Available at: https://beta.slowways.org/ [Accessed 12 July 2021].

Trubina, E. (2020). Sidewalk fix, elite maneuvering and improvement sensibilities: The urban improvement campaign in Moscow. *Journal of Transport Geography*, 83. doi: 10.1016/j.jtrangeo.2020.102655.

TVK (2015). Parkway E40 Bruxelles. Press Release [Online]. Available at: http://cms.tvk.fr/site/assets/files/1997/tvk_dp_bruxellese40-fr-web-1-1.pdf [Accessed 12 July 2021].

TVK, TN, Nivet, S., Ville, O., Fleury, A., Texier-Rideau, G., Ballan, É., ON, RR&A, Franck Boutté Consultants, Yes We Camp, Transamo (2019). *Places du Grand Paris. Principes de conception pour les espaces publics du Grand Paris Express*. Société du Grand Paris, Île-de-France Mobilités, Paris.

Valderrama Pineda, A.F. and Vogel, N. (2014). Transitioning to a low carbon society? The case of personal transportation and urban form in Copenhagen: 1947 to the present. *Transfers*, 4(2), 4–22.

Valegeas, F. (2014). Concevoir et habiter un quartier dit durable : injonctions écologiques et dynamiques collectives à Beauregard (Rennes) et Les Brichères (Auxerre). PhD Thesis, Université Paris-Est, Créteil.

6

The Revival of Active Modes in Global North Cities

Jean-Paul HUBERT[1], Kaduna-Eve DEMAILLY[2] and Clément DUSONG[1]

[1]*LVMT, Université Gustave Eiffel, Marne-la-Vallée, France*

[2]*LADYSS, Université Paris-VIII-Vincennes-Saint-Denis, Saint-Denis, France*

6.1. Introduction: extracting active mobility from the places assigned by functionalist planning

There is something of a political slogan in the term "active mobility". It refers mainly to journeys made on foot or by bicycle in the city, on the scale of a few neighborhoods, or as a complement to journeys by public transportation, for utilitarian (work, study, shopping, etc.) or recreational purposes (walking, tourism, etc.). But the expression also carries a positive connotation that distinguishes this way of traveling from others, because it meets several objectives of public interest in a perspective of moderation for the city and mobility: it mobilizes the whole body and is healthy; it is not in need of a fossil fuel whose combustion emits polluting or greenhouse gases; its infrastructures take up little space; and moreover, it does not create a deadly threat to other road users, who are heavier, faster and better protected. It is also known as "soft mobility" when the frame of reference is no longer that of individual capacity, but rather of quality of life and environmental impacts. The practice of active mobilities is not limited to public roads, although it finds emblematic places in certain major avenues, boulevards, promenades, ramparts, docks, "boards", *ramblas*, etc., where people take pleasure in walking, watching and being seen. It also unfolds in public spaces that to a lesser or greater

Geographical Places in Transportation,
coordinated by Jean-Baptiste FRÉTIGNY.

degree are open, landscaped or commercial: forecourts, squares, gardens, parks, shopping centers, malls, etc., prohibited to motor vehicles, indeed strictly pedestrian-only. Sometimes these places form part of a coherent urban whole, like the historical neighborhoods of many European cities. But, more often, they exist independently of others and are difficult to access for pedestrians and cyclists who are not local residents, especially when these places have been designed and planned to attract users over a large scale. When they were built in open countryside, like the hypermarkets and peri-urban parks and leisure bases built in the 1970s–1980s, their access was designed solely with automobiles in mind (Schut 2017).

With the transformation of the city in the 20th century, usage of active modes has come to be increasingly associated with the commercial, recreational and tourist functions of places. Their role in the general system of urban mobility has thus been progressively marginalized. Although from the 1970s separate networks of cycle paths began to be introduced in the Netherlands, then in Flanders, Denmark and western Germany – such as in Münster in Westphalia – cities have generally been handed over to automobile circulation and the right to circulate by bicycle has gradually been reduced to merely being tolerated on the street network. The pedestrianization of certain streets or shopping districts, carried out in the 1960s–1980s in North American and European cities, simply applied the functionalist principles of separate functions and traffic, so much so that it in fact contributed to the diffusion of automobiles and the technicalization of road management (Beaudet and Wolff 2012). The introduction of "pedestrian streets", emblematic of a new definition of urban centrality (Feriel 2022), therefore went hand in hand, in a context of urban sprawl, with the increase in the capacity of the fast road network and the automobile parking bays enabling access to it. This resulted in more insecurity for local pedestrians and cyclists, and detours in the routes they take (Heran 2011). In the face of this legacy, current policies aim to give more freedom of movement to vulnerable users, pedestrians and cyclists, and better access to the city in general. Restoring political reach for active modes, under the banner of well-being in the city, presupposes redeploying them in the space, from the places where they are practiced, by redeveloping them, further connecting them and ensuring peaceful coexistence between users.

This is the perspective that is examined in this chapter, with the understanding that such objectives are only achievable if road safety, universal accessibility and public transportation policies address the vulnerability of pedestrians and cyclists on streets, the obstacles encountered by people with reduced mobility and the cost of using rapid means of transportation to access urban resources across a metropolitan area. We will focus particularly on western cities, and more broadly the cities of the North, which have experienced the "urban transition" from the pedestrian city to the automobile city (Wiel 1999). In these cities, the promotion of active mobilities involves not only reserved spaces but also a new sharing of the street to allow users of active modes to travel along it more easily and with more continuity.

In this context, we will see that active mobilities were first engaged in a city environmental and health policy, where the rise of their practice became a political objective. This shift requires national regulations, urban planning, the construction and exploitation of statistics, evaluation of public action and mobilization of actors. It has also led to the issues around walking being aligned with those of cycling, despite the interests of pedestrians and cyclists not necessarily converging.

Section 6.3 examines the consequences of the development of active mobility and the need to reserve more room for it in the public space. As we will see below, this question covers several scales, and it encourages a renewal of many representations and models: from those in metropolitan areas to those of places of pedestrian and cyclist circulation, streets and networks, not to mention places of access to public transportation, which are necessary partners of active mobilities. We will conclude on the special relationship between active mobility and inhabitants of the city.

6.2. Words and watchwords of active mobility

Cycling and walking are today considered together under the name "active modes", broadly contrasting with motorized modes of travel. Non-electric skates, skateboards and scooters also come under this category, as do e-bikes, despite being partially motorized, since no additional energy – or "assistance" – is provided without a minimum amount of effort from the cyclist. This is not the case for personal mobility devices (PMD) such as e-scooters and segways.

Would such an association have been proposed at the turn of the 20th century when bicycles and other mechanical devices burst into the public space? The difficult coexistence of cyclists with usual road users presented the populations and public authorities with questions similar to those recently raised by the massive development of PMDs (Bertho-Lavenir 1999). By bringing active modes back into the discourse and narratives that give them meaning, this section discusses what is presupposed by their designation and the new proximity between walking and cycling under the banner of mobility.

6.2.1. *A mobility called to the rescue of major causes*

Cycling resurfaced in North American and European public debate in the 1970s. Discussions emphasized its low energy consumption, but did not address its effects on physical condition, health or well-being. Austro-American philosopher and theologian Ivan Illich's flagship work, *Energy and Equity* (1975), criticizes the counterproductivity of the automobile system, detailing the environmental, social and energy problems generated by the massive motorization of society. Two decades later, other works show the negative consequences of this motorization on health, especially for children, due to a reduction in daily physical exercise and an increase in ambient air pollution (Pucher and Buehler 2010). Walking arises in turn in the debate. In the wake of the 1996 French LAURE (Law on Air and Rational Use of Energy) act, the terms "soft mobility" and "clean mobility" appeared in France to describe non-polluting modes of transportation. The discourse on "clean mobility" then became reinvested in the fight against climate change, taking into account the carbon footprint of transportation. At around the same time, the terms "active modes" and "active mobility" became widespread in health and lifestyle studies. In the French-speaking world, they first appeared in Quebec (Papon 2012). Studies on the health benefits of physical activity demonstrated the value of daily journeys on foot or by bicycle in reducing the risk of obesity and cardiovascular disease (Papon 2002; Celis-Morales et al. 2017).

The adjectives "active" and "soft"[1] might refer to different issues, but in designating the same field of mobility, they become more or less interchangeable. Today, the expression "active modes" has proved most popular, probably for semantic reasons: a positive connotation, associated with the vital value *par excellence*, compared to a connotation of fragility in a domain where exposure to accident risk is always a major concern. These mobility qualifications find institutional recognition in France, notably through multi-year energy programming, designed to achieve the objectives of the Paris Agreements at COP 21. One of its incarnations is the clean mobility development strategy, which, among the "targets, actions engaged and to be undertaken", sets a modal share[2] objective to be achieved for "soft modes" (French Ministry of Ecological Transition 2016, p. 50). The role of directing and evaluating public policies is perceived here through statistical indicators; this will be explored later. This display of a national policy, particularly in the direction of cities, aims to make mobility a lever for action to reorient lifestyles in order to limit the prevalence of obesity, ambient air pollution and CO_2 emissions in the atmosphere.

---

1 This overlap is also found in other countries: *mobilité douce ou active, mobilidad blanda o activa, sanfte oder aktive Mobilität...* in French, Spanish or German speaking countries.

2 Proportion of trips made using a given mode of transportation, excluding trips more than 100 km from home.

6.2.2. *National strategy and local implementations*

Policy measures aiming to increase the practice of active modes are diverse. Some are addressed directly at households through subsidies. In Belgium, the Netherlands and, more recently, France, those who cycle to work can receive compensation from their employer or the State. In addition to this are the *Coups de Pouce* ("Helping Hand") initiatives offered by States or local governments, limited in time, to subsidize the repair of bicycles, parking and the purchase of e-bikes[3].

National policy is also put into practice through territorial planning, subject to regulations, prepared at local government level by the technical services and voted on by elected officials. This is why a significant component of the national strategy consists of developing the legal framework, toward greater freedom of circulation being given to pedestrians[4], cyclists and possibly drivers of electric vehicles. The French "Street use code", "Streets for all" and "Action plan for active mobility" programs, again inspired by examples from Belgium or the Netherlands, have increased the freedom of movement of vulnerable users in low-speed areas[5]: *zones 30* (zones limited to 30 kph) and even more so *zones de rencontre* ("meeting zones"), limited to 20 kph, where motorists must give way to cyclists and pedestrians (see "30 zone" and "meeting zone" notices in Demailly et al. (2021)).

The initiative to embark, or not, on a strategy to promote active modes rests with local authorities by publishing a specific programming document. Pioneering cities in this field are Copenhagen, which stood out in 1962 by applying the principles of architect-urban planner Jan Gehl (Capelli and Chardonnet-Darmaillacq 2019), and, for walking more specifically, Geneva, which in 1995 proposed pedestrian routes to discover the city and get around it. In 2009, San Francisco launched the Pavements to Parks program, which aims to develop, notably for pedestrians and cyclists, parcels of pavement taken from one or more automobile parking spaces, called *parklets*. A general format is emerging, at least in Europe, where these documents are becoming one of the pillars of a general mobility scheme covering the whole of the administered territory, such as the Walkable City plan in Stockholm, adopted in 2010, or the pedestrian maps of the Brussels-Capital Region and the Strasbourg *Eurométropole* in 2012. In an approach to mobility as a system integrating all modes of travel ("systemic and intermodal"), these specific plans program developments or

3 See: www.coupdepoucevelo.fr [Accessed 25 May 2021].

4 To which, from a legal point of view, are assimilated persons in wheelchairs and children on bicycles.

5 Their safety depends first of all on the speed of motor vehicles: the risk of a pedestrian dying in the event of a collision with a car is divided by six between a street limited to 30 km/h and another limited to 50 km/h (Martin and Wu 2015).

engineering structures to reinforce safety at crossroads, improving comfort of practice and restoring pedestrian and cycling frameworks. They present the objective of better sharing public space between modes and defusing conflicts between pedestrians and cyclists (Strasbourg.eu 2012).

Lastly, public action in favor of active mobility applies on the scale of urbanism interventions. These are generally intended to last, but also to be mentioned are the temporary and often more innovative developments, introduced within frameworks qualified as tactical or experimental urbanism (see tactical urbanism notice in Demailly et al. (2021)). As such, the Covid-19 health crisis that began in 2020 gave several countries an additional impetus in the promotion of cycling in cities, in order to direct some public-transportation users toward this mode, deterred by the decrease in supply and the fear of being contaminated. This led to the construction within a few weeks of several hundred kilometers of temporary cycle ways, known as "pop-up bicycle lanes" (Buehler and Pucher 2021).

6.2.3. *Quantification and measurement of objectives*

Policy and public debate are fed by quantitative assessments that measure developments and the distance between them and the objectives. But these models also condition their representations and have a retroactive influence on policies, as Alain Desrosières has shown (2014 for example). In this perspective, confusion regarding the objectives can be noted in the 2016 French strategy "for the development of clean mobility". It questions the degree of attention paid by the State to the specificities of walking and cycling at that time. In indicating that "the share of soft modes (walking and cycling), which in 2008 was 2.7% [...] aims to reach 12.5% in 2030", the French Ministry of Ecological Transition (2016, p. 50, our translation) was in fact quoting the cycling share alone, as had been measured by the National Transport and Travel Survey (ENTD: *Enquête Nationale Transports et Déplacements*) in 2008, and clinically envisaged it multiplying by five, or almost, in just over 20 years[6].

6 In a notice dated May 5, 2017, the French National Council for Ecological Transition (CNTE 2017) proposes to correct the objective to 25% for walking and 12.5% for cycling in 2030. The INSEE-SDES survey on personal mobility in 2019, which continues the ENTD of 2008, mentions a stable share of cycling at 2.6% in 2019 and a share of walking increased from 22.3% to 23.9% (Babet and Trévien 2021). The third envisioned version of the Clean mobility development strategy, published in 2024, removes both the reference to "soft mobilities" and the quantified assessment of modal shift objectives towards active mobility (see Gouvernement français 2024).

In accordance with national strategy, local documents relating to mobilities (urban travel plans or mobility plans) set objectives for reducing the share of car journeys made by people traveling on their own (solo-driving), and increasing the share of travel via public transportation, walking and cycling. They therefore presuppose the existence of means and tools for observing mobility for the monitoring of these programs.

Knowledge of the share of active modes is based on the general observation device for urban mobility: surveys of how the residents of a metropolitan area and its outskirts travel, conducted according to a unified methodology that earned them the name "Cerema certified mobility surveys" (EMC2: *enquêtes mobilité certifiées Cerema*). Most major French cities are subject to a quantitative description of how their inhabitants travel, including on foot or by bicycle, every 10 years[7]. Other evaluations complement this device in order to know, for example, the expectations of pedestrians or cyclists in terms of development, or to appreciate the ease of practicing active mobility, also known as "walkability" or "cyclability", of different places in the city. Over shorter periods, the evolution of bicycle usage can also be measured using data from mobile applications or devices for counting cyclists' flows on selected road sections. From a statistical point of view, these measures are difficult to make representative, but they are beginning to be integrated into mobility observatories to monitor daily traffic trends[8]. Pedestrian counting techniques are also being developed from analyses of video recordings, but they are still only rarely disseminated.

In France, associations of cyclists, and more recently pedestrians, organize periodic consultations on residents' satisfaction with regard to conditions of travel on foot and by bicycle in their municipality[9]. Respondents are recruited on the Internet through associative sites without the guarantee of statistical representativeness; they

7 The interpretation of the modal share figures in these surveys must take into account the fact that, to be counted on foot or by bike, a journey must be made entirely by this mode. However, active modes are widely used in addition to another mode or in places where one mode connects to another (intermodality). According to the EMC2, we walk on average between 250 and 300 m when travelling by bus or tram and 350 and 450 m, respectively, for a journey by metro and train (Rabaud 2015).

8 See, for example, in Île-de-France: www.omnil.fr [Accessed 27 May 2021]

9 See: www.ffrandonnee.fr/la-federation/marche-en-ville/resultats-du-barometre-des-villes-marchables et www.parlons-velo.fr [Accessed 4 October 2021].

nevertheless express an authorized and often expert opinion on the quality of places for the practice of active modes in French cities.

Other quantitative indicators of the family of walkability scores (Duncan et al. 2016) can play a role in defining active mobility objectives. Their calculation incorporates the number of activities accessible on foot from a given place and thus quantifies the degree to which a place's potential is favorable to the practice of walking. Not surprisingly, this kind of indicator highlights dense fabrics, built according to a pedestrian metric, often bourgeois or gentrified. It is of interest to real-estate investors, but it can be criticized for applying a utilitarian conception where pedestrians and cyclists are considered primarily as potential consumers of everyday or cultural goods (Bertoncello et al. 2013). It is essential to develop new, creative and scientifically controlled quantification methods that are more in line with the issues of quality of life, the environment and usage of public spaces (Barry and Keane 2019).

6.2.4. *A combination of different strengths*

Mobility surveys have made it possible to associate walking and cycling under the theme of mobility. This source confirmed the dynamics of active modes from the early 2000s in some cities, while usage is still declining at the national level (Héran 2014; INDDIGO et al. 2020). It lends itself to the expression of political objectives and also makes it possible to differentiate cities. Statistics show the far greater weight of walking compared to cycling, but this significance ratio is reversed in the public and political discourse on active mobility.

The average modal share of walking (see note 10) generally reflects the demographic weight of the densest part of the metropolitan area (where inhabitants walk the most) within the scope of the survey. The percentage of journeys made on foot therefore varies significantly according to the EMC2 surveys and their spatial structure: between 15% (Bayonne and the Basque-Landes coast in 2010) and 39% (Paris region in 2010–2011) for surveys conducted between 2008 and 2018[10].

The median of the 50 surveys conducted during this period is 26%. The cycling modal share is much lower, with a median of 1.5% of trips and less significant

10 Cerema data grouping 50 surveys, used in their widest scope (authors' processing). These shares may differ from the values communicated, which are often higher due to being measured over a narrower and densely populated scope.

variations, but a few cities stand out significantly: Strasbourg[11] (5.6%), Grenoble (3.9%), Bordeaux (3.3%) and Angers (3.1%) in metropolitan France, as well as Cayenne (4.1%) in overseas France. These shares remain significantly lower than those measured in cities in the Netherlands, Denmark, Japan or Germany, countries where cycling is a much more widespread practice (Buehler and Pucher 2012).

Measured in this way, walking and, to a lesser extent, cycling practices follow logics of proximity, that is to say, carrying out activities a short distance from a base location, generally the home. The statistics of an intercommunality such as Nantes Métropole (2015) illustrate the orders of magnitude commonly observed for walking or cycling on a weekday: on foot, an average journey takes 11 min for a distance of 700 m, and by bicycle 14 min and 2.7 km. Cycling enables a speed to be reached that is comparable to public transportation.

Travel on foot or by bicycle to one's place of work or study, for retail purposes (the most common reason) or for other services is generally short in range. A specificity of on-foot mode is the significant proportion attributed to simply going for a walk: about one in 10 journeys on foot are carried out for this purpose, five times more than simply going for a ride with mechanized modes (Rabaud 2015). In France walking tends to be more of a female-dominated practice, whether traveling entirely on foot, or walking in the context of using public transportation. The use of cycling is much more a male-dominated pursuit, almost three times more in the city[12].

Recourse to walking is much more frequent than to cycling, but the activist discourse in favor of active modes is mainly carried by numerous cycling associations (Riou and Dusong 2020), who make consistent claims and are able to rely on an abundance of scientific content produced (Pucher and Buchler 2017), much more so than for walking. These two mobilities are thus represented in the public debate by systems of actors of unequal political strength, which we cannot fail to relate to the much more cycling-oriented content of State plans and strategies, which are nonetheless presented as "active modes" or "walking and cycling". In the spring of 2021, the website of the French Ministry of Ecological Transition,

---

11 Strasbourg is the French city where cycling is most developed and the first to have considered, in 2013, a metropolitan-wide cycling network (Vélostras network project), followed in 2017 by Grenoble (Chronovélo network), Paris in 2019 (VIF network, ex-RER vélo) and Nantes in 2021.

12 Estimate calculated by the authors of the modal shares in urban areas with 50,000 inhabitants or more, according to the 2019 personal mobility survey (SDES, INSEE data). The modal shares of walking alone, public transportation and cycling are, respectively, for women, 31.1%, 15.2% and 1.4% and, for men, 27.3%, 11.2% and 4.1% (individuals aged 6 and over).

dedicated to cycling and walking (Figure 6.1), illustrated at the time the difficulty on an institutional level of having walking exist alongside cycling in national strategies: the measures listed on the page related only to cycling.

Unlike cycling, and hiking, walking in the city does not seem to create as much of a sense of belonging to a collective (Cefaï 2007). With a few exceptions, activism in favor of walking operates through other issues: neighborhoods, heritage, comfort, health and above all the safety of walkers, children, the elderly and people with reduced mobility, in particular. Fighting road insecurity is the most common and universal theme of demands made on behalf of pedestrians. It can be led by associations or national federations (such as the *Place aux Piétons* collective in France, at the initiative of the "walkable cities" barometer mentioned above), international NGOs (Box 6.1) or by local residents' associations, or even media personalities. Of note among them is the Mexican, Peatónito, who dresses as a masked wrestler to jump out into the middle of cars in the congested streets of Mexico City and berate drivers who are not respecting pedestrians (De Llano 2013).

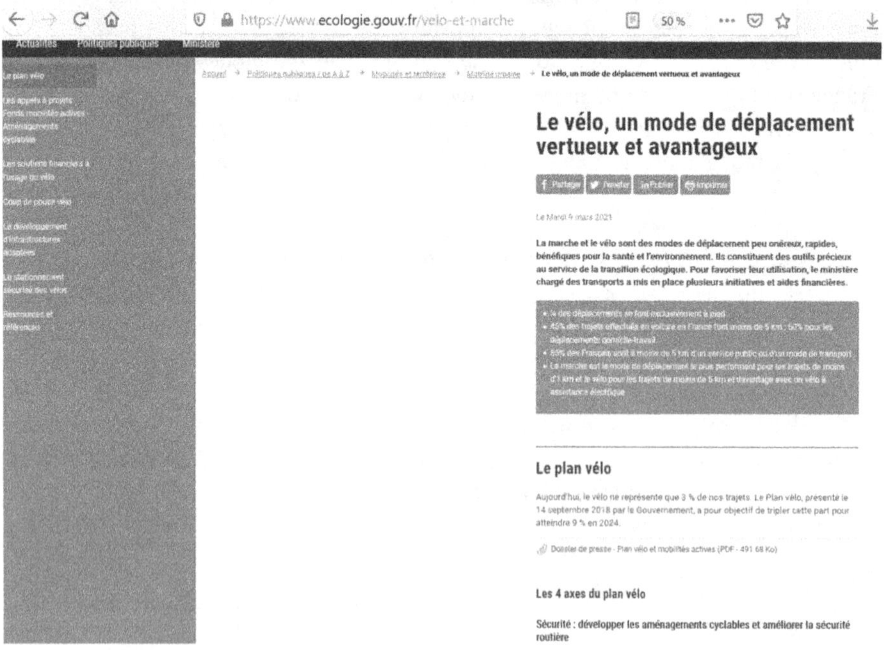

Figure 6.1. *Screenshot of the page "vélo et marche" ("cycling and walking") from the website of the French Ministry of Ecological Transition, May 2021. The menu on the left hand side guides users to projects, cycling projects, infrastructure, secure bike storage, etc.*

In this fragmented landscape, pedestrian activism often aligns with that of cyclists, whether within the framework of common associations (*2 pieds 2 roues*, in Toulouse, or the *ADTC - Se déplacer autrement*, in Grenoble) or in the parallel implementation of "walkability" and "cycling" barometers. One risk would be to reduce the issues of walking in the city to questions of mobility and road calming, and forgetting the actual presence of pedestrians, whether static (sat on benches, terraces, etc.) or active (playing a game, for example), which is so important for places to be animated. The different components of active mobilities are nevertheless united in their need for a new sharing of roads and even a reorganization of relationships between central and peripheral places, as discussed in the next section.

Walk21 is an international non-governmental organization for the promotion of walking. It is dedicated to guaranteeing a "right" to walk for all and to favoring the walkability of cities. It brings together mainly operational actors of walking: political and technical actors of local governments, professional actors and experts as well as associations involved at different levels.

Its activity is divided into five main components:

– the organization of an annual 5-day international conference since 2000 (the first conference was held in London). The cities hosting the conference are generally chosen for their policy and innovative pedestrian achievements;

– the construction of a network of actors committed to pooling thinking and knowledge;

– the promotion of the "international walking charter" formalized at the Melbourne conference in 2006, mainly targeting the development of public spaces in favor of walking and the inclusion of all pedestrian populations;

– the production of resources (guides) to assist in the implementation of public pedestrian policies;

– targeted actions such as the Global Sidewalk Challenge, which aims to build or rehabilitate 100,000 km of sidewalks around the world by 2030.

Box 6.1. *The activities of the international NGO Walk21 (according to www.walk21.com)*

6.3. The sharing of public space in question

The active-mode development policies that we examine have been devised against a backdrop of great difficulty in extending the surface area of circulation places in general. While urban growth was accompanied by the multiplying of infrastructures for motorists, active modes seemed able to make do with narrow road widths or relatively discreet passages to cross large parcels, but with the projected increase in their practice this assessment will need to be reviewed.

Figure 6.2. *Active mobility in its places. Maisons-Alfort (Val-de-Marne, Paris region), April 2019. Photograph by Julie Scapino (MUP, French urban pedestrian mobilities group). For a color version of this figure, see www.iste.co.uk/fretigny/transportation.zip.*

Until the 2000s, bicycles were often required to snake through traffic. As for pedestrians, as Jan Gehl (2010, p. 123) notes of western cities, “it is impressive to note how many obstacles and difficulties have been incorporated into pedestrian landscapes over the years”. This section therefore addresses the conditions for a new sharing of circulation places. The problem is complex and involves global thinking on several scales, as the additional road space needed for active modes must be partly taken from automobile circulation, and the coexistence of pedestrians and cyclists also needs to be regulated (Figure 6.2). Certain spectacular operations have demonstrated how much the automobile has taken over certain places. Thus, the destruction of road viaducts in Nanterre in 2007 (Avenue Joliot-Curie) or in Montreal in 2016 (Bonaventure Expressway) returned places to the city, and not only for active modes. In general, this involves removing a traffic lane or parking lane in order to widen the sidewalks and install cycle lanes, along with public benches or cafe and restaurant terraces. This space taken from automobiles can lead to a general decrease in exchanges and interactions, damaging at the scale of the urban area, hence the significance of travel planning. Part of the challenge is therefore to develop the models of representations of the city as well as those of the street.

6.3.1. *Reintegrating of active modes into general circulation*

Dualistic center/outskirts perceptions elicit the idea of homogeneous outskirts where one can only circulate by car (Demoli 2021). They could therefore be an obstacle to paying sufficient attention to active modes in the secondary hubs and beyond. In today's metropolitan context, centrality is measured by the quantity and quality of accessible places. It depends less on proximity to a geometric center and more on good connections to transportation routes through suburban, agricultural or woodland areas presenting a low population density. This (re)distribution of centers is expressed by the metaphor of the archipelago, a term first applied to the scale of the world-city system, then taken up in the context of polycentric urban areas, such as Rennes (Chapuis 2008). Reconsidered in an archipelago model at the scale of a metropolitan area, or even at a more encompassing scale, active mobilities may well be practiced in a specific manner in certain areas, but they must also allow these areas to be accessible and linked together, with the possibility of recourse to public transportation.

The city of Strasbourg also uses this metaphor to present the active mode development areas in its 2012–2020 travel plan (Chtourbine 2013) (Figure 6.3). The map applies the archipelago model, highlighting areas where the proportion of roadway dedicated to pedestrians exceeds 30%. The pedestrian plan proposes the creation of a pedestrian network described as "masterful", arranged to connect "neighborhood centralities" under the best walking conditions.

Actions in favor of active mobilities, associated with local travel, are aimed first and foremost at a local effect such as the comfort of residents or the securing of places frequented, particularly in the vicinity of schools or public transportation hubs. But it is necessary to place them in a spatial reference framework that goes beyond the local scale, like the archipelago model, to take advantage of the power to connect places through these mobilities. The establishment of shared zones or spaces and especially 30 kph zones illustrates the question posed to collectivities by the spatial scale of interventions in favor of active mobilities. The answers are contrasting according to the territories, the political commitments of elected officials and the modes of decision-making.

Figure 6.3. *Strasbourg, "archipelago city". Areas where the share of the roadway dedicated to pedestrians exceeds 30% (source: RR&A, pedestrian map of the city of Strasbourg (2009))*

Thus, in France, the city of Strasbourg had to abandon plans to extend a 30 kph zone to the entire city center following a referendum in 2011. However, in 2016 the intercommunality of Grenoble extended the speed limit of 30 kph to almost all its roads, with the exception of main highways (with high circulation), limited to 50 kph. In a more homogeneous and dense territory, the City of Paris applied a similar decision in 2021, aligning it with most major European metropolitan areas (Amsterdam, Brussels, London, Madrid, etc.), but not without a variety of objections being raised. It seems significant that, to oppose such decisions, the Automobile Club Association asserts that "the concept of 30 kph zones is intended to improve road safety *in very specific and very limited places*, and not to become a new generalized urban standard" (our translation; emphasis added)[13]. Such a position shows that the change of scale of spaces favorable to active mobilities is far from obvious for certain actors.

6.3.2. *Enhancing the continuity of pathways*

To counter a logic that confines active modes at a local level, it is important to improve the continuity and linearity of pathways. Coordinated action on infrastructure and regulation must allow walking and cycling practices to be extended little by little. Proponents of the usage of active modes thus claim, for pedestrians, cyclists and the like, not only more space to circulate but also priority in all situations involving an intersection or permitted mixing of traffic. Changes to traffic regulations partly respond to this (see above) and are reflected in improvements to roads.

More and more cities are offering cyclists circulation rights and priorities, reserved lanes and parking spaces and facilities. For many years, the struggle of cycling associations has also focused on the consolidation of a continuous, high-capacity bicycle network that is included in a planning document at the scale of the metropolitan area. In the 1990s, the *Réseau Vert* ("green network") of inner Paris was a pedestrian and cyclist scheme designed by activists[14] and its routes were connected through small streets or sidewalks. Adopted by the municipal authorities, the bicycle network has become the legitimate medium for new tourist or logistical usages, open to new vehicles: scooters, cargo bikes, bicycle sharing services, etc. The

---

13 See: www.automobile-club.org/espace-presse/communiques/projet-de-generalisation-des-zone-30-a-paris-l-automobile-club-association-aca-s-y-oppose-et-appelle-a-une-consultation-generale [Accessed 25 September 2022].

14 See: http://reseau.vert.free.fr/association/presentation.html [Accessed 3 December 2021].

construction of this network went hand in hand with that of a network of public and private actors, including an activity of services for cyclists: bicycle rental and repair, secure parking, marking and registration of frames and even cycling training. The express bicycle network project (VIF network, formerly *RER vélo*), formed by 12 routes across 720 km of cycle paths planned between 2025 and 2030 in the Paris region, is nearing realization thanks to the consultation of user associations, communities at different territorial levels (municipal, intermunicipal and regional). It is also proof that this implementation of cycling infrastructure today exceeds the limits of the city center (Dusong 2021).

Similar ideas are developing with regard to walking. The Brussels pedestrian plan states that "the continuity of the sidewalk must be guaranteed, including at crossroads. The pedestrian remains on their pathway, and it is the vehicle that crosses this pathway" (*Région de Bruxelles-Capitale* 2011, p. 46, our translation). The City of Strasbourg is planning, as we have seen, a "masterful" inter-neighborhood network. But the idea of a pedestrian infrastructure is more difficult to conceive of, since pedestrians' usage of the street is not limited to mobility, on the one hand, and traveling on foot is not confined to roads, on the other hand. Such an infrastructure can be broken down into three main subsystems (Hubert et al. 2021) associated with specific places and spaces of circulation. The most important, dedicated to functional travel, is constituted by roads, with the components that are sidewalks, road crossings, squares and forecourts, managed in relation to other traffic. The second subsystem concerns public spaces whose management is not linked primarily to mobility issues (parks and gardens, river banks, cemeteries, etc.) and depends on various administrations. Lastly, the third subsystem depends above all on private actors, who sometimes have very little involvement in maintenance and surveillance: it concerns passages inside large housing estates and buildings, or that cross derelict urban sites (vacant land or wasteland), where walkers define their place for themselves. The high heterogeneity of this infrastructure makes its design complex as a whole. But this notion of infrastructure provides a useful regulatory idea to mobilize the actors that are to be involved in development and accessibility approaches aimed at pedestrians.

6.3.3. *Addressing intermodality with public transportation and coexistence between active modes*

On a larger scale, the development of intermodality places takes on great significance for active modes, which, it must be remembered, are inseparable from public transportation. Public transportation users are very often required to walk or remain standing during connections, on platforms or in vehicles, in conditions of comfort or discomfort that matter to users and operators of the networks alike. Open air connections are sometimes encouraged as a substitute for a journey between two stations on an overcrowded bus or tram line. Thus, Keolis, the operator of the Bordeaux metropolitan transportation network, explained in 2014 to users transiting through the center of Bordeaux that it was in their interest, and that of the network in general, to walk between Hôtel de Ville, Mériadeck and Gambetta stations rather than taking the tram to make the connection (Bosdecher 2014).

Lastly, public transportation requires modes that feed to places of origin or destination, some at a greater distance than others. The use of bicycles combined with another mode of transportation, however, is hampered by this vehicle's poor portability, and safe parking being difficult to find. Bicycle parking is a regular subject of complaint by cycling associations and may become a component of the mobility offer made by public transportation operators. Manufacturers of cycles and equipment, along with vehicle-sharing rental companies, also find room for innovation by offering products adapted to different circulation places and urban contexts. The folding bike thus offers a means of expanding the possibilities of intermodality. The portability of PMDs, similar to that of folding bikes, may well explain in part their success.

In the Netherlands and Japan, intermodality is assured thanks to high-capacity parking lots. These countries stand out with bicycle types that are emblematic and bulky, but adapted to the circulation conditions. The Dutch have tall, wide, comfortable bicycles. In their cities with wide reserved cycle paths, the usage of imposing cargo bikes has developed since the 2000s, facilitating the transporting of young children (Weirich 2012). In Japan, meanwhile, specific cycling facilities are scarcer, forcing cyclists to ride on the roadway together with motorized traffic or on the sidewalk with pedestrians. Japanese *mamachari* bicycles (literally: "mother's bicycles") are equipped with one or two child seats, but they are narrow, to be able to weave through traffic more easily (Steele 2012).

At street scale, local governments are also faced with the choice of installing cycle lanes either on roadways or on sidewalks. One option consists of directing bicycles along bus lanes, or along paths taken up on sidewalks, thus minimizing the share of the roadway taken away from automobiles. Applied in the city center, this pro-car option awakens old conflicts between cyclists and pedestrians, especially in countries where cycling almost disappeared from cities post-1960s. Indeed, pedestrians have become accustomed to finding their bearings and crossing based on what they can hear, and cyclists or PMD drivers approaching noiselessly presents a factor of discomfort and danger, particularly for people who are visually impaired or have reduced mobility. In addition, the trajectory of walkers is difficult for cyclists to predict. The option that tends to prevail today therefore consists of having cyclists circulate on the roadway, which is developed and relieved by reducing speed limits, instead of encroaching on the space reserved for pedestrians (bicycle notice in Demailly et al. (2021)).

6.4. Conclusion: active modes as vectors for the activation of places

Wherever we have encountered or used the expression "active mobility", the word "active" was understood as an adjective affixed to "mobility", considered an attribute of individuals. But it can also be understood in French as the verb "to activate", with mobility the subject. The expression can take at least three meanings. First, as physical exercise, traveling on foot and by bicycle activates the body, hence their place in health policies. Next, this mobility activates emotions and affects, both negative and positive. They are caused by more frequent and diverse interactions with other human and non-human elements of the urban environment, and also result from paying greater attention to all forms of the landscape. Lastly, as the major vector of face-to-face social interactions, these journeys animate public places, recreational and commercial, insofar as one property of these places is that it is open to them. This is why these modes of travel are more often associated with going for a walk, strolling and recreation, but also generate fears of accidents, conflictual confrontations with other pedestrians or cyclists, and even attacks (Ravensberg et al. 2020). Residents and local authorities thus pay great attention to the behavior of users of this mobility and to the standards that define the "good pedestrian" or the "cycling citizen" (Middleton 2018), especially when these users are identified as coming from outside the neighborhood[15]. Thus, walking and cycling activate not only bodies and minds, but also the social and urban environment of individuals. The promotion of active modes in western cities is still confronted with the urban fragmentation that has resulted from a functionalist urbanism based on automobile mobility. The confinement of walking and cycling to a few types of places, central historical neighborhoods or sites with a commercial or leisure function, has the same cause as the lack of physical exercise, road congestion and its dangerousness for city dwellers. Active mobility policy now takes these interdependencies more into account. Despite the difficulty of developing precise assessment tools covering its entire field, it has taken its place as a pillar of a global policy against this urban fragmentation and its social and environmental consequences.

15 Mariano Perelman (2019) shows the importance of this judgment in the testimonies of Buenos Aires rag-pickers working under the often hostile gaze of residents.

6.5. References

Babet, C. and Trévien, C. (2021). Marcher et pédaler : les pratiques des Français. *Datalab Essentiel* [Online]. Available at: https://www.statistiques.developpement-durable.gouv.fr/marcher-et-pedaler-les-pratiques-des-francais [Accessed 17 March 2022].

Barry, K. and Keane, J. (2019). *Creative Measures of the Anthropocene. Art, Mobilities and Participatory Geographies*. Palgrave Macmillan, Singapore.

Beaudet, G. and Wolff, P. (2012). La circulation, la ville et l'urbanisme : de la technicisation des transports au concept de mobilité. *VertigO*. doi: 10.4000/vertigo.11703.

Bertho-Lavenir, C. (1999). *La roue et le stylo. Comment nous sommes devenus touristes*. Odile Jacob, Paris.

Bertoncello, B., Hernandez, F., Bertoni, A. (2013). L'accueil du piéton dans les espaces publics du centre-ville de Marseille : discours, pratiques, actions. *Recherche Transport Sécurité*, 29, 279–293.

Bosdecher, L. (2014). Bordeaux : les usagers des transports en commun incités à se déplacer à pied. *Sud-Ouest* [Online]. Available at: https://www.sudouest.fr/2014/08/08/bordeaux-les-usagers-des-transports-en-commun-incites-a-se-deplacer-a-pied-1637253-2780.php [Accessed 3 December 2021].

Buehler, R. and Pucher, J. (2012). Walking and cycling in Western Europe and the United States: Trends, policies, and lessons. In *Transportation Research Board Annual Meeting*. TR News, Washington, D.C. [Online]. Available at: https://trid.trb.org/view/1143635 [Accessed 29 November 2021].

Buehler, R. and Pucher, J. (2021). COVID-19 impacts on cycling, 2019–2020. *Transport Reviews*, 41(4), 393–400. doi: 10.1080/01441647.2021.1914900.

Capelli, C. and Chardonnet-Darmaillacq, S. (2019). Entretien avec Jan Gehl : There is much more to walking than walking. *Espaces et sociétés*, 179, 17–40. doi: 10.3917/esp.179.0017.

Cefaï, D. (2007). *Pourquoi se mobilise-t-on ? Les théories de l'action collective*. La Découverte, Paris.

Celis-Morales, C.A., Lyall, D.M., Welsh, P., Anderson, J., Steell, L., Guo, Y., Maldonado, R., Mackay, D.F., Pell, J.P., Sattar, N., Gill, J.M.R. et al. (2017). Association between active commuting and incident cardiovascular disease, cancer, and mortality: Prospective cohort study. *BMJ*, 357, 1456. doi: 10.1136/bmj.j1456.

Chapuis, J.-Y. (2008). Rennes : La Ville archipel et son corollaire : La Ville des proximités. *Annales des Mines – Responsabilité et environnement*, 4(4), 37–43. doi: 10.3917/re.052.0037.

Chtourbine, E. (2013). Comment réaliser un plan piéton pour une ville ? Le Plan Piétons de la Ville de Strasbourg. In *Colloque : Le Piéton au cœur de la ville*. CEREMA, Strasbourg [Online]. Available at: https://www.cerema.fr/fr/system/files/documents/2017/11/e_Chtourbine_plan_pieton_Strasbourg_cle2d3389.pdf [Accessed 29 November 2021].

Demailly, K.-E., Monnet, J., Scapino, J., Deraeve, S. (eds) (2021). *Dictionnaire pluriel de la marche en ville*. Editions de l'Œil d'Or, Paris.

Demoli, Y. (ed.). (2021). *Peut-on se passer de la voiture hors des centres urbains ?* MSH Paris-Saclay Éditions, Gif-sur-Yvette.

Desrosières, A. (2014). *Prouver et gouverner. Une analyse politique des statistiques publiques*. La Découverte, Paris.

Duncan, D.T., Méline, J., Kestens, Y., Day, K., Elbel, B., Trasande, L., Chaix, B. (2016). Walk score, transportation mode choice, and walking among French adults: A GPS, accelerometer, and mobility survey study. *International Journal of Environmental Research and Public Health*, 13(6). doi: 10.3390/ijerph13060611.

Dusong, C. (2021). Les dynamiques de l'usage du vélo dans les espaces de banlieue : continuités, décalages et ruptures. Le cas de l'Île-de-France. PhD Thesis, Université Paris-Est, Marne-la-Vallée.

Feriel, C. (2022). *La ville piétonne. Une autre histoire urbaine du XXe siècle ?* Éditions de la Sorbonne, Paris.

Gehl J. (2010). *Cities for people*. Island Press, Washington, D.C., Covelo, London.

Gouvernement français (2024). Projet de stratégie de développement des mobilités propres [Online]. Available at: https://concertation-strategie-energie-climat.gouv.fr/sites/default/files/2024-11/SDMP_vF.pdf [Accessed 13 March 2025].

Héran, F. (2011). *La ville morcelée : effets de coupure en milieu urbain*. Economica, Paris.

Héran, F. (2014). *Le retour de la bicyclette*. La Découverte, Paris.

Hubert, J.-P., Monnet, J., Scapino, J. (2021). Infrastructure pédestre : une notion pour agir [Online]. Available at: https://www.construction21.org/france/articles/h/dossier-mobilites-18-infrastructure-pedestre-une-notion-pour-agir.html [Accessed 29 November 2021].

Illich, I. (1975). *Énergie et équité*. Le Seuil, Paris.

INDDIGO, VERTIGOLAB, ADEME (2020). Impact économique et potentiel de développement des usages du vélo en France. Report, Direction générale des entreprises (DGE), Paris [Online]. Available at: https://librairie.ademe.fr/mobilite-et-transport/332-impact-economique-et-potentiel-de-developpement-des-usages-du-velo-en-france-en-2020.html [Accessed 29 November 2021].

de Llano, P. (2013). Un vengeur masqué contre les chauffards de Mexico. *Courrier international* [Online]. Available at: https://www.courrierinternational.com/article/2013/08/15/un-vengeur-masque-contre-les-chauffards-de-mexico [Accessed 27 May 2021].

Martin, J.-L. and Wu, D. (2015). Accidentologie des piétons. Report, IFSTTAR, Marne-la-Vallée.

Middleton, J. (2018). The socialities of everyday urban walking and the "right to the city". *Urban Studies*, 55, 296–315. doi: 10.1177/0042098016649325.

Ministère de la Transition écologique (2016). Stratégie de développement de la mobilité propre [Online]. Available at: https://www.ecologie.gouv.fr/sites/default/files/Strat%C3%A9gie%20d%C3%A9veloppement%20mobilit%C3%A9%20propre.pdf [Accessed 29 November 2021].

Nantes Métropole (2015). Temps, distances et temporalités des déplacements [Online]. Available at: https://auran.org/sites/default/files/publications/documents/06_temps_et_distance.pdf [Accessed 27 May 2021].

Papon, F. (2002). La marche et le vélo : quels bilans économiques pour l'individu et la collectivité ? Deuxième partie : la sante et la sécurité. *Transports*, 413, 187–197.

Papon, F. (2012). Le retour du vélo comme mode de déplacement. HDR Thesis, Université Paris-Est, Marne-la-Vallée.

Perelman, M. (2019). La marche dans la production de l'inégalité sociale. La collecte informelle des déchets à Buenos Aires. *Espaces et sociétés*, 179, 145–160. doi: 10.3917/esp.179.0145.

Pucher, J. and Buehler, R. (2010). Walking and cycling for healthy cities. *Built Environment*, 36(4), 391–414.

Pucher, J. and Buehler, R. (2017). Cycling towards a more sustainable transport future. *Transport Reviews*, 6(37), 689–694. doi: 10.1080/01441647.2017.1340234.

Rabaud, M. (2015). La mobilité à pied en milieu urbain. In *Journée : Une Voirie Pour Tous, Observer et évaluer les modes actifs*. CEREMA, Lille [Online]. Available at: http://voiriepourtous.cerema.fr/IMG/pdf/3_Voirie_pour_tous_10-02-2015_La_Marche_Urbaine_cle249368.pdf [Accessed 29 November 2021].

Ravensbergen, L., Buliung, R., Laliberté, N. (2020). Fear of cycling: Social, spatial, and temporal dimensions. *Journal of Transport Geography*, 87, 102813. doi: 10.1016/j.jtrangeo.2020.102813.

Région de Bruxelles-Capitale (2011). IRIS 2. Plan de mobilité de la Région de Bruxelles-Capitale. Bruxelles Mobilité-AED, Brussels.

Riou, D. and Dusong, C. (2020). Le rôle des associations dans le "retour du vélo" en Ile-de-France. *Note Rapide Mobilité*, 834 [Online]. Available at: https://www.institutparisregion.fr/nos-travaux/publications/le-role-des-associations-dans-le-retour-du-velo-en-ile-de-france/ [Accessed 3 December 2021].

Schut, P.-O. (2017). Habiter la ville par le sport et les loisirs ? La création des bases de plein air et de loisirs (1964–1976). *Leisure/Loisir*, 41(1), 47–68. doi: 10.1080/14927713.2017.1337522.

Steele, M.W. (2012). The making of a bicycle nation: Japan. *Transfers*, 2(2), 70–94. doi: 10.3167/trans.2012.020206.

Strasbourg.eu (2012). Plan piéton ville de Strasbourg 2011-2020 [Online]. Available at: https://fussverkehr.ch/wordpress/wp-content/uploads/2016/09/plan-pieton.pdf [Accessed 29 November 2021].

Weirich, C. (2012). Cargobikes – The solution to urban congestion, current usage, future potential and impacts of an alternative way of short distance transportation. Master's Thesis, University of Hasselt, Hasselt.

Wiel, M. (1999). *La transition urbaine ou le passage de la ville pédestre à la ville motorisée.* Mardaga, Sprimont.

7

Sensory Urban Mobilities: Experiences and Uses of Ambiances

Damien MASSON
PLACES, CY Cergy Paris Université, Cergy-Pontoise, France

7.1. Introduction

Bus stops, stations and airports form numerous movement-places (Amar 1989) that link urban public spaces to the lines, tubes and flows of transportation networks. These places are designed to ensure the co-presence of passengers, facilitate their circulation and make access to their means of travel possible. As much as they provide the technical infrastructure for urban mobility, vehicles and transportation spaces also provide a material framework in which each individual movement is lived as an ordinary experience of mobility.

They involve bodies that accelerate, slow down, stop, brush against or touch others, often involuntarily and without any desire to do so. Forward-looking gazes, facing toward where the most effective movement is occurring, contrast with others, more haggard-looking, confused, seeking an indication or a sign to orient them. Droplets running down people's faces suggest perspiration perceptible down to the poles of the subway. The Covid-19 pandemic provides a reminder, from under surgical masks, of the breaths and the life of the mouths that breathe, talk, cough or fog up glasses. It has also revealed the extent to which daily mobilities contribute to

Geographical Places in Transportation,
coordinated by Jean-Baptiste FRÉTIGNY.

the sensory environment of urbanized worlds: sound, light, heat, smell, etc. The multiple lockdowns across the world in 2020 and 2021 resulted in the disappearance of a resonance nevertheless consubstantial with urban life, made up of the metabolic and continuous audible layer[1], or drone, of the sounds of motors, of rolling, squealing and air masses split by vehicles. The marks of this disappearance – and the reappearance of the presence of animals – have been widely documented[2] and heritagized via maps[3] of these new sound environments.

Beyond documenting all the sensory effects of transportation (or their absence) in urban environments, this chapter shows the importance of taking into consideration ordinary sensory experiences of urban mobilities in public transportation. They assume specific material and technical assemblies, sites and networks, which participate in singular sensory environments. In addition to their utilitarian logics, transportation places provide opportunities for encounters and exchanges and engage affects as well as emotions.

In order to study these different dimensions, the ambiance approach seems particularly appropriate. Understanding of the material contexts and the sensory dimension in which social life unfolds, ambiances allow situations to be described as social interactions by relating them to places – when we evoke a *station or airport ambiance* – or to the activity that characterizes them – a *beehive ambiance*, for example, which refers metaphorically to situations where a compact and mobile crowd occupies subways during rush hour, as is visible in Figure 7.1. Ambiances also reflect the point of view of individuals with regard to these situations, their emotional or affective involvement: a *reassuring atmosphere* or a *hostile climate*, for example.

A pragmatic[4] approach to ambiances makes it possible to illuminate what is at play in public transportation places. It involves showing, on the one hand, that they are not simply the result of material contexts and social situations, because they are also the subject of institutional desires and technical engineering. On the other hand,

---

1 The metamorphosis sound effect (in French "*métabole*") describes "the unstable and changing relations between elements of a sound ensemble" (Augoyard and Torgue 2006, p. 73). In other words, it refers to charged sound environments, composed of a multiplicity of constantly changing sources.

2 As evidenced by the proliferation of sound recordings and videos on the Internet – on social networks, video portals, like ad hoc sites – giving something to hear and see of silent and empty streets.

3 See, for example, one of the most complete: https://citiesandmemory.com/covid19-sounds/.

4 That is, interested in the question of action and in the ability of individuals to adjust to situations.

they participate in the multiple sets of rules – formal and tacit – of transportation[5] users by exercising power over them.

The foundations of sensory urban research and the specificity of the ambiance approach form the subject of section 7.2. Focusing on the definition of ambiances and their linking to mobilities, it shows in particular how this research perspective contributes to the renewing of questions relating to transportation and urban mobilities by the qualification of the sensory, social and affective dimensions of individuals' experiences, which fosters the development of original methodological approaches. The subsequent sections take a closer look at the uses that are made of ambiances in urban transportation places. Focusing on users, section 7.3 examines in what way their actions contribute to the formation of ambiances. Section 7.4 looks at the production of ambiances by operators, following the practical challenges associated with transportation production. Section 7.5, which focuses on security issues, illustrates how the sensory environment and ambiances are instrumentalized as a form of soft power to influence bodies and police conduct.

Figure 7.1. *Compact crowd and bustle at Montparnasse-Bienvenüe station during rush hour (source: Masson (2011)). For a color version of this figure, see www.iste.co.uk/fretigny/transportation.zip*

5 For ease of reference, "transportation" refers here to all public transportation infrastructures accessible to the public: vehicles, platforms allowing access to them, stations and subspaces that make them up (rooms, galleries, corridors, etc.) and their thresholds in the urban space.

7.2. Ambiances and mobilities, ambiances of mobilities: specificities of an object and of an approach for places and transportation

It may be tempting to affirm that each transportation technique corresponds to an ambiance, in the same way that Marc Desportes indicates that "each major transport technology models […] an original approach to the space one passes through, each major technology carries its own "landscape" within it (2005, p. 8, our translation). Transportation places often evoke strong sensory universes, whether they are marked with the seal of the exceptional for many users, such as certain major stations and major airports, with vast, luminous spaces (Figure 7.2), where the audio announcements fall from on high, which they send out in contrast to the everyday bustle, in tunnels, some somber, some well-lit, clean, with aromas composed of mechanical, electrical and organic odors from the subways and underground trains. However, focusing only on the material and sensory characteristics of spaces to qualify their ambiances would lead to forgetting that they are the product of a relationship with those who experience these spaces, subjects that are neither neutral nor passive.

Figure 7.2. *The grandiose architecture of London St. Pancras station, all in volumes, modenature and ornaments (source: Masson (2012)). For a color version of this figure, see www.iste.co.uk/fretigny/transportation.zip*

7.2.1. *Architectural and urban ambiances?*

In everyday vocabulary, ambiance refers to a double reality. Defined in the French dictionary *Le Petit Robert* as "the moral and material atmosphere surrounding a person" (our translation), ambiance characterizes material conditions (physical or architectural) as sensory, on the one hand, and the climate (political, zeitgeist, or state of mind) of the situations experienced, on the other hand. More broadly, it is used in social sciences to designate "people's sensory potential" (Kazig and Masson 2015), particularly in its spatial dimension. Initially developed in France in the field of architectural and urban research, the works on ambiances qualify the material and sensory environment of social practices and explore the role played by sensoriality in their daily realization. Thus, at play there is an ecology of the body in its built environment. In the paradigm of architectural and urban ambiances, the ambiance "puts us in contact with the whole situation and therefore proceeds from an ecological approach to perception" (Thibaud 2003, p. 88). This ecological conception also makes it possible to point out the mediating role of ambiances: they are a gateway to the world; we feel, perceive and act through them.

This understanding of ambiance comes close to the notion of *Atmosphäre*, frequently used in the German literature, which is characterized by "the way in which it communicates a feeling to us as participating subjects" (Böhme 2013). Böhme, a philosopher, thus clearly places ambiance in an intermediate position, between the subject and the object. He also insists on the active dimension of subjects, as well as on the ability of ambiance to convey meaning. This author also assumes an aesthetic approach (reception and production) to comprehend ambiances. It is thus a question of considering them as productions, staging, that can be felt, that can be conceived as well as perceived, which Jean-François Augoyard (2004) formulates elsewhere. By qualifying the relationships between individuals and the sensory environment, these approaches focus on what is inherent to ambiances, even if to put in parentheses the "moral" dimension evoked in the definition given in *Le Petit Robert*. This void is filled by the concept of affective atmospheres, developed in English-language geography (McCormack 2008; Anderson 2009; Bissell 2010). This includes in its theorization of atmospheres dimensions that escape the world of phenomenon through the notion of affect. Affects express the capacities of bodies to open up to others and to the world (Pile 2010). Transpersonal affects pass through us and can be communicated between people. We can illustrate with walking within a crowd, which implies an ability to be affected by the crowd while our walking influences those of the individuals walking around us. The works on affective atmospheres thus highlight the ways in which the affects and the intensities of interindividual relationships that they sketch out are also configurators of ambiances. These studies show how the emotional dimension of our experiences touches us and therefore helps to give a tone, a color or a temperament to ambiances such as they are lived. *In fine*, we can draw from ambiances that they are there all the time, all around individuals. They are the subject of an experience and are a way for researchers to qualify situations. They derive from social conducts, usages of the body and related affective intensities.

Three modes of understanding ambiances make sense for the study of the sensory dimension of ordinary mobilities. Ecological allows an insistence on the configuration of sensations, perceptions and practices by ambiances. The aesthetic approach, for its part, questions ambiances as intermediate products, between the intention of conception and staging of a place – likely to be mobilized by operators of transportation spaces – and its ordinary aesthetic reception – as a judgment issued by users. Lastly, entry through the notion of affective atmospheres, used in the English-language field, opens the focus by integrating into the field of study of ambiances dimensions related to affects and emotions, which go beyond the sensory domain but contribute significantly to the experience of places.

7.2.2. *Ambiances and mobilities: a methodological porosity*

These modes of comprehending ambiances make it possible to envisage distinct themes of study without them being contradictory, because they share a common interest in the sensory experience of places. To study this, researchers had to develop original methodological tools. One consequence, by no means insignificant, of this necessity is that the notion of ambiance is found to designate both an object (ambiances) and an approach (through ambiances). The works produced in the French-language field of architectural and urban ambiances merge at the intersection of architectural design, engineering sciences (thermal, acoustic, lighting, etc.) and micro-scale approaches to social interactions, particularly originating from symbolic interactionism and ethnomethodology[6]. Consequently, the specificity of the ambiance approach is that it involves constant attention to three registers of observed situations: their materiality, sensory phenomena present and social interactions.

These works develop methodologies situated halfway between experimental approaches and sensory ethnography, which articulate sensory readings, of spaces and bodies (bodyographies[7]), notations of movement (choreographies), measurements of physical phenomena, audiovisual recordings, situational speech

6 Two microsociological trends that focus particularly on social interactions and the ordinary conduct of individuals. In the strand of symbolic interactionism, emphasis is put on the place of sign and symbol systems (such as language) to give meaning to interactions. The ethnomethodological perspective, meanwhile, notably focuses its attention on routine activities to account for them in order to understand how they are used rationally by individuals. The methodological consequences of these approaches notably consist of close attention being paid to verbal practices, such as the description or production of reports, of everything that is observable or that can be said of situations.

7 This notion refers to the idea that urban experiences form part of bodies. Hence, bodyographies designate "a kind of body mapping, where the mapped object cannot be distinguished from its representation" (Dultra Britto and Berenstein Jacques 2012, author's translation).

production, etc. The multiplicity of these approaches has a twofold objective. First, it is a question of ensuring a form of methodological triangulation consisting of comparing observations resulting from different methods and validating their interpretation. Second, this multiplicity creates gateways to the unsayable and the inexpressible, the pre-reflexive and the incorporated, and to the fleeting or even the "aphenomenal" (Balez 2020). To try to achieve this, these approaches involve the almost systematic experience of situations by favoring the immersion and mobilization of bodies, in other words, the methodological use of bodily motion. Because "motion, and by extension, mobility is therefore a *sine qua non* condition of the atmosphere" (Kazig et al. 2017), moving – in all its forms and scales – proves constitutive of ambiances just as much as it allows their understanding. This explains the wealth of the method of commented walks (Thibaud 2001), which – like its multiple representations: sound, onboard and sensory routes, alone or with others, etc. – counts as one of the most widely used within this field, as is the analogous method of "go-alongs" (Kusenbach 2003), particularly widely used in the English-speaking world.

The attempt to access phenomena, impressions or sensations that sometimes escape the visible and conscious through the setting in motion that they engage is also constitutive of "mobile methods" (Urry 2007; Büscher et al. 2010). These methods, which aim to extend the range of tools existing in social sciences, came to be labeled "mobile" in the second half of the 2000s to mark their place in the furrow carved by the mobilities paradigm (Sheller and Urry 2006), which claims a mobile turn in social sciences. Sociologists, anthropologists and geographers involved in this line of research call for methodologies that can grasp what is specific to mobilities. According to Law and Urry (2004), these specificities are due to non-causal, complex, distributed, labile, floating processes that "slip and slide between one place and another" and take "different shapes in different places". Mobile methods aim to understand processes that, like ambiances, are intangible and ephemeral, that appear at the threshold of representations and give prominence to sensory approaches as well as to "non-representational methodologies"[8] (Vannini 2015). Especially documenting experiences and practices, mobile methods favor

8 This term, coined by Vannini (2015), refers to a broad and composite set of research methods and techniques borrowing from human geography, as well as from material anthropology, cultural studies, the arts and performance, sociology of the body and of emotions, etc. They all have in common an interest in the "flow… of everyday life" (Thrift 2008). More prosaically, they can aim at an understanding of sensory, temporal and routine phenomena, for example, non-verbal and non-textual relationships with others and the world.

almost exclusively qualitative approaches and encourage the creation of new protocols that involve (among others): the sensory permeation of researchers in their fields; their setting in motion; in situ deployment of tools to capture (video, sound, photo, etc.) and receive or interact (mobile telephones, MP3 players, GPS, video games, etc.).

The methodologies developed by French-language research on ambiances can nevertheless be criticized for two reasons (Masson and Thomas 2020). First, they focus attention on the sensory dimensions of situations while not relating them to the social, political or cultural backgrounds in which they take on meaning. Furthermore, although recording observations in situ, they often do so by means of experimental protocols that can distance it from ordinary experience. On "mobile methods", Merriman (2014) notes that caution should be exercised with regard to methodological creation, which is not to be taken as an issue in itself and involves several risks. The first is the risk of forgetting that existing research on transportation is not limited to quantitative or modeling works. Second is that of considering mobility research as a discipline in its own right in social sciences, having its epistemology and justifying its own methodology. Last, that of forgetting the existence of many works that did not wait for the announcement of a "mobility turn" in order to study ordinary bodily mobilizations as the complex and processual, sensory and ephemeral dimensions of mobilities.

7.3. Users as producers of ambiances: what sensory experiences of ordinary mobile practices?

Unlike the French-language works, which have sought rather to characterize the ambiances of transportation places, the English-language works approaching mobilities by atmospheres have largely directed their attention to the practice and lived experience of users, by varying the modes of transportation, like the modes of experience studied. These last studies make it possible to understand, for example, the relationships of individuals to the world in situations of mobility, construction of identities in co-presence, the micropolitics of encounters such as the grip on the body and the bodily dimension of power games.

Thus, certain works grasp the phases of stasis, the immobile moments of mobilities. Drawing on waiting situations, the works of Bissell (2007) and Vannini (2011) re-examine this time constituting ordinary mobilities. Through sustained observation work and close attention to bodies, they enable a deconstructing of the image of mobility conceived as a temporal performance (of speed) or as a binary alternation between moving and stopping. They show that waiting takes many forms. They also invite us to think of occupying layover times based on a continuous regime of activities (from passive, at least in appearance, to active), which provides opportunities for encounters and exchanges, or indeed can be the object of desire.

Among the works that examine active modes, such as running (Cook et al. 2016) and cycling (Spinney 2010; Jungnickel and Aldred 2014), the mobilization of affective atmospheres enables questioning of the meanings attributed to these bodily practices by those who conduct them. Far from conveying the expected image of cyclists or joggers who would include their practice in a logic based purely on physical performance, as in a logic of distinction from other users of public space, these works show that the construction of the identities and relationships of these athletes to the world is not unique or fixed. They describe how the ambiances associated with these practices are constructed in games of negotiation for sharing space with other users and by the feeling of self-exposure, sometimes perceived critically in view of the danger that their practice poses to them when faced with motor vehicles. Lastly, they account for the creation of sensory configurations specific to these practices, showing that athletes, deconstructing, for example, the image of listening to headphones as forming a bubble, instead place themselves in a phenomenal boundary situation, linking public space and personal sensory sphere.

In addition, much research questions ordinary mobilities in public transportation, subways, trains, shared taxis, etc., from the angle of the specific sensitivities that they orchestrate. Some studies empirically probe the visual experience on trains, for example (Bissell 2009). Others study the constitution of ambiances on urban trains and subways. They show that they are configured to intersect temporalities specific to different sensorial regimes – visual, sound and body practices, as evidenced by the practice of subway musicians, performance that is much more related to the temporal and phenomenal adjustment to the moving sound environment of trains in motion than to the rough addition of a sonority that is completely unaware of its context (Masson 2009). They also highlight that ambiances derive from affects between passengers, their activities and the lesser or greater extent to which they are able to handle others, like their "misanthropic dispositions" arising from a co-presence that is more endured than desired (Bissell 2010). These works challenge the idea that each transportation would have an ambiance that would be its own, stable and always true. On the contrary, they insist on their multiplicity and on their lability, which can change instantaneously, according to modulations of the sensory environment (variation in ambient luminosity, for example) or practices and interactions, as illustrated by the bodies tensing and preparing to disembark from a train as it approaches a stop.

Recent works address more head-on the micropolitical dimensions of the ambiances of ordinary mobilities. They question the management of otherness between passengers whose social identities, uses of the body as sensory expansions – for example, the noise they make, the place they take up, the volatility of the odors of what they may be carrying, etc. – diverge. As Catherine Earl (2018) shows in the case of daily mobilities in Ho Chi Minh City, these differences are reflected in the ability to implement a social distinction through senses by the "mixophobic" elites (Earl 2018). They prefer taxis, even if hampered by heavy rain, in order to avoid sensory exposure, to benefit from a vehicle where temperature and humidity are controlled, as well as to avoid unwanted odors and sounds, associated with captive users of public transportation. These differences are also related to the circulation of affects – tension and anger, anxiety and fear of violence, but also joy or complicity for others. At the scale of bodies, they translate macropolitical relationships of inequality in terms of ethnicity (Eales and Thomas 2021), class (Bissell 2016), age and culture (Wilkinson and Wilkinson 2018) and gender (Martin 2021). These analyses expose how these circulations of affects create ambiances, particularly of discomfort or even threat, which have the effect of leaving certain passengers unable to challenge the inequalities to which they are subjected. For others, these circulations of affects make it possible to create a bond and temporary understandings, even forms of reassurance and increased capacity for action (empowerment).

This diverse set of research illustrates the significance of ambiances in the transportation experience. It highlights the thematic, methodological and theoretical proximities between the studies of affects, ambiances and mobilities, and it emphasizes the active role of passengers in the production of atmospheres. What, now, of operators of transportation spaces and their intentions with regard to producing an ambiance?

7.4. The production of ambiances by the operators of transportation spaces

The material contexts in which actions and encounters between individuals take place are also the subject of intentions relating to design, production and management of their sensory dimensions. Thus, it is common to see explicit references to the notion of ambiance appear, in the programming of urban public space, a certain idea of which – always positive and meliorative – must be reached at the end of the project. Through this notion, issues of quality of life, privileged usages, but also urban identity, comfort, accessibility, etc., can be targeted.

7.4.1. *Forming an ambiance?*

In enclosed spaces, the task of controlling sensory parameters of the material environment is facilitated. Shopping centers, restaurants, hospitals, museums, etc. are subject to sensory shaping, or even staging, as shown in Figure 7.3, by an action on the composition, materials, colors, brightness, and temperature, as well as on the auditory, olfactory, and indeed tactile and podotactile dimensions.

Figure 7.3. *A sound and light installation that radiates the space with its presence in a shopping center in Stockholm (source: Masson (2011)). For a color version of this figure, see www.iste.co.uk/fretigny/transportation.zip*

Not escaping this rule, subway stations, galleries and interconnection spaces are all "hyperconditioned" environments [...], spaces that no longer offer any connection with the (natural, urban) environment in which they are set. Decontextualized, they are thus defined by the break that they impose from the prevailing conditions around them" (Siret and Requena 2019, our translation). More broadly, transportation places such as stations, airports and vehicles are also subject to this hyperconditioning, because access to so-called natural light is subject to fine control of the filtering capacities permitted by glazing. These walls provide mediation of the relationship with the outside – just like the expression of a social and economic distinction made visible by the size of the glazed surfaces and especially the degree of transparency of glass products. This quest to control ambiances is the subject of very distinct engineering and lines of expertise (thermal, acoustic, lighting, etc.), which raise the question of possible synthesis. While the managers of a transportation operator, the contracting authority for a project, can express the desire for a singular ambiance, it will be less the product of a composition of intentions and more a sum of expertise, and financial, technological and normative constraints. Moreover, the prospect of a homogeneous ambiance, often desired, particularly for reasons of image and commercial identity, forms part of an ideal that necessarily faces the test of user practice, which constantly evolves and in turn renews sensory phenomena such as ambient affective intensities.

7.4.2. *What ambiances for transportation places, and in order to do what?*

For mobility operators, the production of ambiances in transportation places is not an end in itself; it must serve practical intentions. The ambiance of the subway, for example, comes from the singular characteristics of its premises. Essentially underground, subways are formally characterized by spaces distinct from the surface, where tubes, vaults, closed rooms and stairs are linked together. The underground environment created with the subway is a sensory world in its own right, which must be understood by its users in order to find their bearings, orient themselves and feel comfortable: to be able to use them serenely, like any other space open to the public. These issues are gaining in significance as station personnel are gradually removed, increasingly replaced by technical devices: signage, maps, ticket vending machines, ticket gates, call boxes, etc. One of the consequences of this evolution is an image of a dehumanized subway (Tillous 2016), which proves problematic in situations of traffic disruption (Atienza and Masson 2011) and also in promoting a feeling of safety[9].

Many works illustrate the interest the RATP places in its sensory space, such as those focusing on the design of the olfactory environment, as of the 1950s (Thiery 2004), on sound (Grosjean 1988) or on the semiotics of its spaces to promote their

---

9 The results of the 2019 edition of the survey by the *Institut Paris Région* (Noble 2020) on the feeling of insecurity on transportation in the Paris region show that the deployment of SNCF and RATP safety agents, along with security forces, are the two most popular measures, according to 59.1% and 45.4% of respondents, respectively, to reinforce the feeling of safety in transportation. Moreover, the 2013 "Living Environment and Security" survey by France's National Institute of Statistics and Economic Studies (INSEE) and the French National Observatory of Crime and Criminal Justice (ONDRP) states that the lack of other passengers in transit is cited as an anxiety-provoking factor for 58% of respondents (Vanier and d'Arbois de Jubainville 2018, p. 11).

legibility and facilitate the directing of users in the 1980s (Peny 1992). They seek action levers to improve comfort, optimize the operation of transportation places and ensure management of passengers' emotions.

Ambiance is also at the heart of the identity issues raised by transportation space operators. With regard to airlines, the work of Lin (2015) demonstrates how Singapore Airlines' commercial identity is closely linked to the ambiance that it aims to procure, which goes beyond the strict framework of the sensory environment. A combination of affects, relating to the friendliness, benevolence and body of the flight attendants (mainly women), contributes to a stereotype of "oriental hospitality" based on an ambiance of "care", where passengers are well looked after. In the case of the Paris subway and RER suburban railway, we have shown (Masson et al. 2012) how the sound ambiance, beyond advertisements and commercial jingles, characterizes the identity of this network. The acoustics of the premises, users and technology co-produce its sound identity. It is plural in form – depending on the places and types of lines – and changes throughout the day. Despite changes in ambiance, we observe a sense of identity permanence – particularly in sound – shaped by modes of usage, such as walking speeds and ways of speaking in public. This permanence is also reinforced by enduring markers, including the sounds of ticket validation signals, the distinctive noise of motorized vehicles and rolling stock, and their reverberation, which varies depending on whether spaces are vaulted or have high ceilings.

From a more prospective point of view, the production of a methodology for characterizing the sound ambiances of crossing, associated with successively transiting from place to place in the subway, can inform the design of future spaces (Masson 2014), such as the stations of the Grand Paris express. Nevertheless, the discourse of the *Société du Grand Paris*, which is responsible for the identity of this new network, indicates that this identity will be more driven by the aesthetics of objects in the station than by the sensory environment (Société du Grand Paris 2015a). In another case, the identity project carried out by the designers of the future line E of the Lyon subway is qualified by the terms of "happy atmospheres" (Rémy and Paxinou 2020). For designers, these ambiances are not accompanied by specific developments, they are not prescriptive of forms and are defined solely with regard to their end purposes in terms of the experience for users. One may wonder about the practical translation of such a proposal. While illustrating a variety of expectations associated with ambiances, these examples reflect the lack of clarity surrounding the latter's production methods, despite numerous attempts. They also testify to a risk of cosmetic usage of ambiances, where the objective of communication prevails in the design imagination and aims to mask the logics of budgetary austerity applied to projects.

Intervention regarding luminosity also constitutes an important field of action for the ambiance of railway stations and the subway. Artificial light is predominant and the desire for brightness has guided the design of the Paris subway from its very beginnings. The choice of beveled white tiling responds to a challenge of maximizing the reflection of light along the surfaces, as shown in Figure 7.4.

Figure 7.4. *White tiling and high luminosity, a paradigm that remains topical. Pont Cardinet subway (line 14, Paris), opened to the public in 2021 (source: Masson (2022)). For a color version of this figure, see www.iste.co.uk/fretigny/transportation.zip*

Since the creation of this subway, we have witnessed a consistent increase in the amount of lighting: the number of lux – the unit of measurement of illuminance – has been multiplied by 40 in the space of a century (Peny 1992). The architectural charter of the new Grand Paris express stations also pursues this objective of high luminosity. However, it adopts a principle consisting of renouncing artificial light as much as possible to ensure "a natural-light continuity between the station hallway and the city outside" (Société du Grand Paris 2020), as its author, Jacques Ferrier, indicates (our translation). The architect Dominique Perrault, in charge of Villejuif-Institut-Gustave-Roussy station, even openly defended the abolition of the dichotomy between surface level and below ground. He calls for porous stations, whose climate is an extension of that of the urban environment:

> The traveler will be in direct contact with the ambient air and cascading natural light. They will not get the sense that they are 50 meters below ground, because they will also be in contact with the ambient temperature: when it is cold, they will feel cold, when it is

> hot, they will feel hot, when it is cold outside, it is cold on the underground, etc. (Société du Grand Paris 2015b, our translation)

We thus seem to be moving away from the logic of hyperconditioning, but for what reasons? Ferrier insists on issues related to sustainable development (reducing as much as possible the use of artificial light) and also legibility – an open station is in principle a station where you can easily find your way. The question can nevertheless be raised as to the quality, in particular audio, of such architectures based on a predominantly visual imagination. The last two stations on line 13 of the Paris subway, Les Agnettes and Les Courtilles, delivered in the late 2000s, are emblematic of this. Built on the paradigm of an open well from the platform to the street, they demonstrate poor sound performance for recent spaces: ambient noise during the arrival of trains, difficulty understanding audio messages, and exposure of local residents to noise nuisance, for example (Masson et al. 2012).

Figure 7.5. *Strong lighting, a contrast between light on the platform and darkness on the tracks and a "Be kind" poster campaign encouraging passengers to be more civic-minded at Paddington Station (Elizabeth line, London), which opened to the public in 2022 (source: Masson (2022)). For a color version of this figure, see www.iste.co.uk/fretigny/transportation.zip*

We can also detect between the lines of these development intentions a renewal of the forms of combating darkness, considered problematic culturally. Western modernity always regulates this through lighting (Edensor and Dunn 2020), as can be seen in Figure 7.5. The latter also serves as a remedy for below ground, the image of which still does not seem to be fully supported by those who are in charge (Liehr and Faÿ 2007). More broadly, these visual issues of clarity, openness, visibility and legibility that are also found in the renovation programs of the Paris subway stations[10] form part of a double paradigm: creating a feeling of safety and seeking to limit crime through urban design, or situational prevention[11].

7.5. Power for ambiances? Safety in question

A crucial question for transportation places and their ambiances concerns the feeling of safety. Note that 60% of Parisians say that on at least one occasion they feared being victims of theft or assault on transportation in 2019 (Noble 2020, p. 14), while 23% of French people say they were victims of theft or an attempted theft, all modes combined, in 2020, including during the Covid-19 lockdown periods (Bernardi 2022, p. 44). These fears are territorially situated: certain lines and places are more frequently cited as anxiety-inducing than others, and they prove gendered, being experienced by women more than by men (Condon et al. 2005). The survey of Parisians indicates that the representations associated with insecurity concern, in descending order based on the number of responses: social transgressions (reference to people who are "under the influence of alcohol", "intoxicated by drugs", or "uncivil"[12]), then the degree of frequentation of transportation places ("deserted" or "crowded") as well as the type of frequenting ("absence of personnel") and lastly, social differentiation ("groups of youths", "people in a situation of exclusion").

The environmental aspects ("narrow recesses", "damage", "dirt", "darkness", or "lack of cameras") each account for less than one-tenth of responses. This last point contrasts with the results of the SNCF "safety barometer" (Vanier and d'Arbois de Jubainville 2018, p. 6), which suggests that a feeling of spaciousness, cleanliness, fluidity or the "lively" character of a station contribute to a feeling of safety. These results indicate above all that the feeling of safety is first and foremost a matter of affects and that the action of operators on the sensory environment contributes only in a secondary manner to the production of a safe ambiance. Operators' interventions in this area are not always innovative. The visual policies mentioned in

---

10 We refer here to the *Renouveau du métro* ("Subway renovation") program led by the RATP in 1998 and the *Renouveau des gares* ("Station renovations") program undertaken by the RATP with *Ile-de-France Mobilités* in 2016.

11 Invented in the 1960s and theorized in the 1980s, situational prevention is a *dispositif* that hypothesizes that the shaping of space and its modes of occupation reduce opportunities for committing criminal acts, through a range of urban design techniques.

12 All the categories in parentheses are translations of the nomenclature used by the *Institut Paris Région* that the respondents were invited to employ. Since these qualifiying terms are, in various respects, to be treated with caution, they are given here in quotation marks to distance the authors from their usage.

the previous section (lighting, openness, etc.) are, for example, part of the range of actions deployed in the Paris region by the IDFM organizing authority for mobilities, the RATP and the SNCF to strengthen users' feeling of safety (Région Île-de-France et al. 2018). While contributing to reinforcing comfort and ease for users, they reconnect with secular formal proposals, such as those of the design of Copenhagen Central Station, aiming to help combat nervousness and anxiety among passengers (Löfgren 2008, p. 338).

More subtle ambient safety work is also underway in transportation places. It is based on a double movement: incorporating affects into the security strategies of operators and reproducing them in the routine actions of users (Adey 2014). They can notably be observed in the uses of "nudge theory". This translates into strategies based on the use of cognitive biases for behavioral purposes, which form just as many nudges. A renowned nudge for the male population is the fly sticker placed inside urinals to encourage users to aim better (and therefore to improve the cleanliness of toilets, as well as optimize cleaning budgets), originating in a transportation place, Amsterdam-Schiphol airport (Mestres 2021). Moreover, there have been an increasing number of works devoted to mobility, particularly in the field of behavioral sciences, as well as that of management. They aim to show how nudges can, for example, lead to an increased preference for public transportation (Franssens et al. 2021) or reduce automobile driving speed (Rubaltelli et al. 2021). Based on experiments with precise parameters, these studies nevertheless do not know anything of the social identities of their participants, their capital or their ordinary interactions and gatherings. Those who put in place nudges further to these studies are therefore largely unaware of their real effects.

Closer to ambiances, we also observe the deployment of certain "environmental nudges" promoting safety in transportation. Relatively common in private parking lots, the playing of classical music through speakers appears to have made its appearance in the Toronto subway, and was subsequently adopted in several western subway and railway station networks, in England, the United States and Australia. It has been the subject of experiments in several stations within the Paris region since 2012 (Vielcanet 2022, p. 61). The usage of instrumental music from the 17th and 18th centuries follows a twofold objective of creating a soothing sound ambiance and excluding a specifically targeted population, as cynically indicated by the director of the Newcastle Metro Network (BBC 1998), the first to adopt this approach in England. Although such remarks would probably no longer be so openly expressed today, the use of classical music as a public policy tool, as a weapon (Thompson 2017), aims just as much at social segregation. The moral imagination associated with this musical style in spaces where transit is the norm tends to lead to the vacating of these areas by people who would otherwise linger there, that is, loiterers.

Loiterers are indirectly designated by the categories "groups of youths" or "excluded individuals" mentioned above, targeted by strategies for "combating vagrancy" (Vielcanet 2022, p. 58, our translations), notably deployed in French stations. Such measures are implemented because these undesirable people with deviant behavior would contribute to creating an "atmosphere of uncertainty, nuisance, threat and idleness" (Dewsbury 2014, p. 432), without being markedly objectionable. In many transportation places in the United Kingdom, operators are pushing the offensive usage of sound further. They deploy mosquitoes, that is, boxes diffusing very high-frequency sounds at high intensity (above 17 kHz) to keep certain groups away. Although illegal under British legislation, the European Convention on Human Rights and the UN International Covenant on Economic, Social and Cultural Rights (Little 2015), these measures insidiously assume the targeting of undesirable social groups because of their age, with younger people being the only ones whose hearing is likely to pick up this ultrasound.

More recently, the "Work and station" facilities, present in the SNCF stations of the Paris region, come to mind, as well as the "*Point connect*" charging stations (Figure 7.6), deployed in the Paris subway and RER suburban railway. They consist of equipping these places with seats provided with tablets and charging sockets for telephones and computers. Would they represent an interruption in the incitement to transit rather than to take a pause?

These installations do not quite break with the functional and normative script of fluidity. They allow stopping, but in places designed for it, leaving room for only a limited number of users – two at most – who – presumably workers – feature among desirable social figures, called upon to strengthen ecological control in places where authority representatives are not always there to do so.

These workstations, in the same way as "mosquitos", classical music or other lighting or coloring instruments, contribute to the array of biosocial techniques aimed at acting on bodies to manage affects. These tools reflect a gradual erasure of the boundary between actions to explicitly combat certain conducts and those in favor of a feeling of safety. They also cross ethical boundaries, through recourse to the acceptable and to a socially situated legitimacy, to the sub-phenomenal, or what is below the limits of the felt. Perceptible to a lesser or greater degree, often little noticed, these measures of questionable effectiveness nevertheless exist and continue to be deployed, especially under the weight of a growing "safe city" industry.

Figure 7.6. *A "Point connect" (on right of picture) at Châtelet-Les Halles station, a stopping place in contact with flows (source: Masson (2022)). For a color version of this figure, see www.iste.co.uk/fretigny/transportation.zip*

7.6. Conclusion

Transportation and its places are eminently sensory environments, loaded with affects. Their operators use the ambiances for meliorative purposes, following logics of aesthetics and image but also comfort, care and orientation likely to influence the conduct of users. We understand that the expectations underlying the production of ambiances, going beyond aesthetic concerns, are of the same order as those associated with situational prevention and "nudges": making of the configuration of places, as well as their sensory characteristics, action levers likely to influence perceptions, representations, affects and the actions of users. Contemporary action programs on transportation and its spaces, however, do not make these intentions of ambiances a primary or autonomous development principle, which illustrates a form of operator lucidity with regard to the expectations of users, but also the powers of ambiance.

Two avenues can be sketched out. The first consists of extending non-visual actions so as to place the sensory development of transportation places in a "more than visual" paradigm. The challenge is to include users more, not only by appealing to certain forms of rationality and cognition, but also by considering them as situated, social bodies. The second avenue invites us to think of individuals more as coproducers of ambiances and affects. The challenge is to better move away from certain scientific works and operational studies that can, even today, be impregnated in various respects with deterministic logics of a stimulus-response type, where any action on the physical or sensory environment would result in standardized types of conduct or predictable emotional responses. It is important to better understand the heterogeneity of conduct, the multiplicity of passengers, as well as the variability of the states of each and every individual. Ambiances are neither a luxury nor an object sculpted in accordance with the wishes of a demiurge. They resonate like a sensory echo of lived situations. Their consideration by urban transportation operators, both to understand the modalities of the experience and to act on places, underlines the whole interest, heuristic as well as practical, of thinking of passengers as literally incarnate, made of flesh and bone. This is undoubtedly a more thorny challenge for development actors today than the many functional considerations that most often attract their attention.

7.7. References

Adey, P. (2014). Security atmospheres or the crystallisation of worlds. *Environment and Planning D: Society and Space*, 32(5), 834–851.

Amar, G. (1989). Lieu-Mouvement : les enjeux de l'évolution des stations de métro. Report, RATP, Département Développement Prospective, Vincennes.

Anderson, B. (2009). Affective atmospheres. *Emotion, Space and Society*, 2(2), 77–81.

Atienza, R. and Masson, D. (2011). Des annonces à l'ambiance : qualification et amélioration des situations de diffusion et de réception des annonces sonores dans le réseau ferré de la RATP. Report, RATP, Département Développement Prospective, Paris, 168.

Augoyard, J-F. (ed.) (2006 [1992]). *Sonic Experience: A Guide to Everyday Sounds*. McGill-Queen's University Press, Montreal.

Augoyard, J.-F. and Torgue, H. (1995). *À l'écoute de l'environnement : répertoire des effets sonores*. Parenthèses, Marseille.

Balez, S. (2020). At the edges of the phenomenal. The design of architectural and urban ambiances as a reconciliation between ourselves and the world. In *Proceedings of the 4th International Congress on Ambiances, Alloaesthesia: Senses, Inventions, Worlds, Réseau International Ambiances* [Online]. Available at: https://hal.archives-ouvertes.fr/hal-03220249.

BBC (1998). Symphonies soothe train vandals [Online]. Available at: http://news.bbc.co.uk/1/hi/uk/51939.stm [Accessed 19 April 2022].

Bernardi, V. (2022). Insécurité et victimation : les enseignements de l'enquête Cadre de vie et sécurité. Report, Service statistique ministériel de la sécurité intérieure, Paris.

Bissell, D. (2007). Animating suspension: Waiting for mobilities. *Mobilities*, 2(2), 277–298.

Bissell, D. (2009). Visualising everyday geographies: Practices of vision through travel-time. *Transactions of the Institute of British Geographers*, 34(1), 42–60.

Bissell, D. (2010). Passenger mobilities: Affective atmospheres and the sociality of public transport. *Environment and Planning D: Society and Space*, 28(2), 270–289.

Bissell, D. (2016). Micropolitics of mobility: Public transport commuting and everyday encounters with forces of enablement and constraint. *Annals of the American Association of Geographers*, 106(2), 394–403.

Böhme, G. (2013). The art of the stage set as a paradigm for an aesthetics of atmospheres. *Ambiances. Environnement sensible, architecture et espace urbain* [Online]. Available at: https://journals.openedition.org/ambiances/315.

Büscher, M., Urry, J., Witchger, K. (eds) (2010). *Mobile Methods*. Routledge, London.

Condon, S., Lieber, M., Maillochon, F. (2005). Insécurité dans les espaces publics : comprendre les peurs féminines. *Revue française de sociologie*, 46(2), 265–294.

Cook, S., Shaw, J., Simpson, P. (2016). Jography: Exploring meanings, experiences and spatialities of recreational road-running. *Mobilities*, 11(5), 744–769.

Desportes, M. (2005). *Paysages en mouvement*. Gallimard, Paris.

Dewsbury, J.-D. (2014). Loiterer. In *The Routledge Handbook of Mobilities*., Adey, P., Bissell, D., Merriman P., and Sheller M (eds). Routledge, London, 429–438.

Dultra Britto, F. and Berenstein Jacques, P. (2012). Corpographies urbaines – Expériences, corps et ambiances. In *Ambiances in Action/Ambiances en acte(s) – International Congress on Ambiances*, Montreal.

Eales, C. and Thomas, H. (2021). (Dis)comfort in the city: How young travellers in London negotiate mobility within the city. *Emotion, Space and Society*, 40, 100814.

Earl, C. (2018). Senses of distinction. Social differentiation, metro-mobilities and daily life in Ho Chi Minh City. In *Senses in Cities: Experiences of Urban Settings*, Low, K. and Kalekin-Fishman, D. (eds). Routledge, London.

Edensor, T. and Dunn, N. (eds) (2020). Introduction: Venturing into the dark: Gloomy multiplicities. In *Rethinking Darkness*. Routledge, London.

Franssens, S., Botchway, E., de Swart, W., Dewitte, S. (2021). Nudging commuters to increase public transport use: A field experiment in Rotterdam. *Frontiers in Psychology* [Online]. Available at: https://www.frontiersin.org/article/10.3389/fpsyg.2021 633865.

Grosjean, M. (1988). Métro : espace sonore. Report, RATP, Paris.

Jungnickel, K. and Aldred, R. (2014). Cycling's sensory strategies: How cyclists mediate their exposure to the urban environment. *Mobilities*, 9(2), 238–255.

Kazig, R. and Masson, D. (2015). L'ambiance comme concept de la géographie culturelle francophone. Défis et perspectives. *Géographie et cultures*, 93–94. doi: 10.4000/gc.3969.

Kazig, R., Masson, D., Thomas, R. (2017). Atmospheres and mobility: An introduction. *Mobile Culture Studies. The Journal*, 3(3), 7–20.

Kusenbach, M. (2003). Street phenomenology: The go-along as ethnographic research tool. *Ethnography*, 4(3), 455–485.

Law, J. and Urry, J. (2004). Enacting the social. *Economy and Society*, 33(3), 390–410.

Liehr, G. and Faÿ, O. (2007). *Les souterrains de Paris : légendes, mystères, contrebandiers, cataphiles.* De Borée, Romagnat.

Lin, W. (2015). "Cabin pressure": Designing affective atmospheres in airline travel. *Transactions of the Institute of British Geographers*, 40(2), 287–299.

Little, C. (2015). The "Mosquito" and the transformation of British public space. *Journal of Youth Studies*, 18(2), 167–182.

Löfgren, O. (2008). Motion and emotion: Learning to be a railway traveller. *Mobilities*, 3(3), 331–351.

Martin, J. H. (2021). Exploring the affective atmospheres of the threat of sexual violence in minibus taxis: The experiences of women commuters in South Africa. *Mobilities*, 17(3), 301–316. doi: 10.1080/17450101.2021.1942171.

Masson, D. (2009). Musique à bord ! Transports collectifs et expression musicale. *Les cahiers du GERHICO*, 13, 75–86.

Masson, D. (2014). Des métropoles sonores ? Quels enjeux pour une lecture, aux yeux fermés, des villes ? *Les cahiers de la métropole*, 70–73.

Masson, D. and Thomas, R. (2020). From a sensitive ecology to a political ecology of ambiances. In *Proceedings of the 4th International Congress on Ambiances, Alloaesthesia: Senses, Inventions, Worlds, Réseau International Ambiances* [Online]. Available at: https://hal.archives-ouvertes.fr/hal-03220314.

Masson, D., Bardyn, J.L, Leroux, M. (2012). Quelle identité sonore pour les espaces de la RATP ? Contribution à une ethnographie sonore du métro parisien. Report, RATP, Paris.

McCormack, D.P. (2008). Engineering affective atmospheres on the moving geographies of the 1897 Andrée expedition. *Cultural Geographies*, 15(4), 413–430.

Merriman, P. (2014). Rethinking mobile methods. *Mobilities*, 9(2), 167–187.

Mestres, A. (2021). Blanc comme Nudge. *Society*, 150, 40–43.

Noble, J. (2020). Sentiment d'insécurité dans les transports franciliens. 2019 survey. Report, Institut Paris Région, Paris.

Peny, A. (1992). Les dimensions sensibles de l'espace transport. *Les Annales de la recherche urbaine*, 57(58), 17–23.

Pile, S. (2010). Emotions and affect in recent human geography. *Transactions of the Institute of British Geographers*, 35(1), 5–20.

Région Île-de-France, Île-de-France Mobilités, SNCF, RATP (2018). Campagne de lutte contre le harcèlement dans les transports. "Ne minimisons jamais le harcèlement sexuel". Press Release, Paris.

Rémy, N. and Paxinou, E. (2020). Happy atmospheres. Metro stations as sound places of happiness. In *Proceedings of the 4th International Congress on Ambiances, Alloaesthesia: Senses, Inventions, Worlds* [Online]. Available at: https://hal.archives-ouvertes.fr/hal-03220337.

Rubaltelli, E., Manicardi, D., Orsini, F., Mulatti, C., Rossi, R., Lotto, L. (2021). How to nudge drivers to reduce speed: The case of the left-digit effect. *Transportation Research Part F: Traffic Psychology and Behaviour*, 78, 259–266.

Sheller, M. and Urry, J. (2006). The new mobilities paradigm. *Environment and Planning A: Economy and Space*, 38(2), 207–226.

Siret, D. and Requena, I. (2019). Architectures des milieux hyper-conditionnés. *Les Cahiers de la recherche architecturale urbaine et paysagère* [Online]. Available at: https://journals.openedition.org/craup/3036.

Société du Grand Paris (2015a). Le design des équipements et mobiliers des futures gares confié à l'agence Patrick Jouin [Online]. Available at: https://www.societedugrandparis.fr/gpe/actualite/le-design-des-equipements-et-mobiliers-des-futures-gares-confie-lagence-patrick-jouin [Accessed 21 February 2022].

Société du Grand Paris (2015b). La future gare Villejuif-Institut Gustave Roussy, vue par son architecte [Online]. Available at: https://www.societedugrandparis.fr/gpe/actualite/la-future-gare-villejuif-institut-gustave-roussy-vue-par-son-architecte-720 [Accessed 21 February 2022].

Société du Grand Paris (2020). Jacques Ferrier : "L'architecture des gares doit respecter le quotidien des voyageurs" [Online]. Available at: https://www.societedugrandparis.fr/gpe/actualite/jacques-ferrier-larchitecture-des-gares-doit-respecter-le-quotidien-des-voyageurs-3120 [Accessed 10 January 2022].

Spinney, J. (2010). Improvising rhythms: Re-reading urban time and space through everyday practices of cycling. In *Geographies of Rhythm*, Edensor, T. (ed.). Ashgate, Aldershot.

Thibaud, J.-P. (2001). La méthode des parcours commentés. In *L'espace urbain en Méthodes*, Grosjean, M. and Thibaud, J.P. (eds). Parenthèses, Marseille.

Thibaud, J.-P. (2003). La ville à l'épreuve des sens. HDR Thesis, Institut d'Urbanisme de Grenoble, Grenoble.

Thompson, M. (2017). To soothe or remove? Affect, revanchism and the weaponized use of classical music. *Communication and the Public*, 2(4), 272–283.

Thrift, N. (2008). *Non Representational Theory: Space, Politics, Affect*. Routledge, London.

Tillous, M. (2016). Le métro comme territoire : à l'articulation entre l'espace public et l'espace familier. *Flux*, 103104(1), 32–43.

Urry, J. (2007). *Mobilities*. Polity Press, Cambridge.

Vanier, C. and d'Arbois de Jubainville, H. (2018). Le sentiment d'insécurité dans les transports en commun : situations anxiogènes et stratégies d'évitement. Report, Grand Angle. Observatoire national de la délinquance et des réponses pénales, Paris.

Vannini, P. (2011). Mind the gap: The tempo rubato of dwelling in lineups. *Mobilities*, 6(2), 273–299.

Vannini, P. (2015). *Non-Representational Methodologies: Re-Envisioning Research*. Routledge, New York.

Vielcanet, A. (2022). Agir sur le sentiment de sécurité dans les transports collectifs franciliens. Report, Institut Paris Région, Paris.

Wilkinson, S. and Wilkinson, C. (2018). Night-life and young people's atmospheric mobilities. *Mobile Culture Studies*, 3, 77–96.

8

Paratransit Places: Spaces of In-Betweenness in Cities of the Global South

Solène BAFFI[1,2]
[1]*Géographie-cités, Université Paris 1 Panthéon-Sorbonne, Paris and Aubervilliers, France*
[2]*Stellenbosch University, South Africa*

8.1. Introduction

In cities of the global South, where paratransit accounts for a large proportion of daily travel, transportation spaces take various forms: they may be developed spaces along colonial train stations or markets, or vacant land lacking equipment where operators, mechanics, scrap dealers and street vendors gather.

This diversity of configurations was highlighted during the Covid-19 crisis: although paratransit operators were quick to apply barrier measures inside vehicles (accepting a smaller number of passengers, distributing hand sanitizer to those boarding the vehicle), observance of social distancing within stations and locations where operators gather was sometimes more complex to implement. In Kigali (Rwanda), ground markings facilitated user compliance with health regulations. On the contrary, in Dakar, queues of users got longer as a direct result of drivers applying a lower occupancy rate. Likewise in Nairobi, where tensions would erupt as passengers rushed to the vehicles (Delaunay and Baffi 2020).

Geographical Places in Transportation,
coordinated by Jean-Baptiste FRÉTIGNY.

These different examples from cities in sub-Saharan Africa testify to the laboratory role played by paratransit places in developing cities. Their high frequentation, including during the Covid-19 epidemic, make them central locations in cities of the Global South.

But paratransit places[1] are also places of resistance, where the imposing of exogenous rules and standards such as those set out by the World Health Organization clashes with the reality and daily practices of local actors. The introduction of certain health measures also offers a reminder of the adaptive nature of these self-organized transportation services, which operate within a context of political, economic and social change on an international scale (Steck 2004). Paratransit places can thus be likened to spatial, temporal and normative "in-betweenness": as anchor places in changing metropolitan areas, they reveal the differentiated logics of the actors at different scales and prove to be representative of the transition processes that animate cities of the Global South.

Thus, a heuristic reading of paratransit places is proposed in this chapter, considering them as privileged observatories for understanding societies and territories with all the complexities presented by cities of the Global South. Paratransit places are considered more particularly in the light of African metropolitan areas where these services today still often constitute the main public transportation supply, if not the only service provided.

In order to grasp the heuristic scope of paratransit places, three frames of reference are proposed: beyond sectoral anchoring, transportation and its places are considered through the prism of urban, political and social geography. Above all, it is the fundamental role of transportation places that is presented, whether at the scale of the paratransit sector itself, city practices or often-fragmented metropolitan areas. Then, stations and assembly points are analyzed as places of power in the urban space, sources of tension and appropriation by different groups of actors. Lastly, paratransit places as sites of affirmation of urban change are examined: the introduction of new models and new technologies reveals deeper socioeconomic transformations, representative of the transitions experienced by cities of the Global South.

1 In this chapter, the expression "transportation places" refers to the places where transportation occurs (stations, vehicles, etc.) in contrast to "transportation nodes/spaces", which refers to a broader approach, covering all transportation-related places (associated commercial functions, public spaces, etc.).

8.2. Places that shape metropolitan areas

Paratransit places are a structuring element of this economic sector, urban life and the growth of metropolitan areas.

8.2.1. *Paratransit, a sector structured by its places*

In most cities of the Global South, the transportation provision is largely reliant on private operators, which offer services on one-size-fits-all networks or according to on-demand transportation operation, with a wide variety of modes of operation. These transportation services have emerged spontaneously in contexts where the institutional transportation offer does not exist, or where it is lacking. They have largely developed since the second half of the 20th century, to the point of today ensuring the bulk of travel in many cities of the South, by means of public transportation (minibuses) or semi-public transportation (tricycles and motorcycle taxis). Despite the diversity of urban systems, modes of transportation and services, certain common characteristics justify their designation under a common label in the English- and French-language scientific literature. Thus, in 1987, Xavier Godard proposed the notion of "transport artisanal" to highlight the atomization of vehicle ownership in the sector and the autonomy of independent operators. Other francophone authors seek to describe the transportation offer provided instead, such as Lammoglia (2013), who speaks of "adaptive" transportation. In the English-language literature, the term *paratransit* seeks primarily to describe a very flexible transportation offer, which is not determined by fixed schedules and can be combined with a higher capacity transportation offer (Salazar-Ferro 2015). Whether they relate to operators' mode of functioning or to the type of service provided, these definitions highlight the fundamental aspects of the sector and its organization. They contrast with the lack of precision, or even the normative and ethnocentrist conception of the expression "informal transport", which would define them as being the negative of formal transport (Godard and Teurnier 1992; Wester 2018).

The organization of the sector and its operating rationale are well observed at the scale of stations and parking lots. Given the logic of atomization of the paratransit sector, internal competition is particularly strong. Indeed, in a context where the unemployment rate remains high, the existence of low barriers to entry into the paratransit sector explains that it turned out to be a "safe sector" for many people. Operators thus organize themselves above all to define and enforce the rules that determine the movements of vehicles: for this, they gather in the same place, thus creating stations or stopping points (Godard and Teurnier 1992). For example, in N'Djamena, Chad, motorcycle-taxi operators spontaneously gather in the urban space, under the supervision of professional organizations that control access to assembly points, and verify the affiliation of drivers and whether they are permitted to access them (Heinrichs et al. 2016). In general, these professional organizations determine the routes of each, the order of departure of the vehicles and the passenger distribution, and they authorize drivers to return to the station at the end of the trip in exchange for payment of a flat fee. These organizations also provide protection for operators, whether in relations with users or with local authorities, such as the motorcycle-taxi unions in Lomé (Guézéré 2012).

Moreover, the fundamental role of stations and assembly points also makes them anchor points and landmarks for actors in the sector. Indeed, a significant number of actors gravitate in and around the paratransit sector: in the case of collective modes, drivers are often assisted by conductors, who collect fares from passengers and ensure that the journey runs smoothly. "Coxers" – playing the role of dispatchers – also assist the crews to ensure the distribution of passengers between the different vehicles within stations or vehicle (un)loading spaces. These individuals, whose activity could be considered marginal, have gradually asserted themselves as indispensable actors in the provision of services (Seck 2006) and their function is intrinsically linked to the space of the station. For many of these individuals, the station constitutes a gateway, providing a first job for migrants arriving in the city, or even a stepping stone (Lombard et al. 2004). In addition to representatives of professional organizations (whose head offices are often located within stations), vehicle maintenance employees and mechanics gather in these spaces to sell their services on a daily basis. Beyond these different actors whose activity relates directly to the transportation service, other peripheral activities flourish at stations or stops, especially small-scale retail.

From the place where drivers gather, whether it takes the form of a stop, a station or a parking lot, the sector is thus subject to formalization, which distinguishes it from services operated illegally. In South Africa, for example, this differentiation leads to the dissociation of, on the one hand, the "shared taxis" type of transportation that exists in townships[2], *amaphelas*, and, on the other hand, paratransit services recognized and regulated by the authorities, black taxis or minibuses. *Amaphelas* are clandestine taxis that circulate along erratic routes, at the request of the customer, which explains their name, which literally means "cockroach", by analogy with the irregular pace at which drivers progress. In his observation of the activity of *amaphelas* operators, Bradley Rink (2022) states that they travel across the township

2 Townships were originally housing lots planned for African populations. They have become more broadly African and mixed-race neighborhoods, a change linked to the evolutions in urban policies within the context of apartheid. This word refers to a very specific urban landscape, in contrast to other less formal forms of occupation of urban space, such as slums, for example.

space without pausing, in the absence of an anchor point where they can stop. Within the continuum that exists between informal and formal activity, including clandestine transportation services, motorcycle-taxi services (taxi-type service) and collective minibus services (service on a one-size-fits-all network), the typology of the stopping points, operator gathering places and stations offers landmarks.

8.2.2. *Paratransit stations and stops: places of resources and sociability*

For city dwellers, transportation places represent veritable urban resources: as places of centrality, they catalyze activities and see the weaving of social ties. In the cities of the South, particularly in the cities of sub-Saharan Africa, a significant proportion of the population lives in conditions of economic and social precariousness, which are often materialized by spatial relegation to outer suburban neighborhoods (Chaléard et al. 2014). Accessibility to urban resources (places of employment, shopping, training, etc.) is then identified as a tool for social and spatial inclusion and a lever to fight poverty (Farrington 2007; Bertrand 2010; Paquette 2010; Fol and Gallez 2013), and transportation places can be seen as central places to access urban opportunities. In addition to the access to networks that they offer within a context of spatial relegation, transportation places are often sites where activities are concentrated. Transportation and trade remain closely linked activities, and stations and stops are privileged places for merchants to set up (Monnet 2007). Street vendors make the most of passenger waiting times to offer their goods and services. These business practices evolve in parallel with the feminization of employment: while they are often responsible for household purchases and meal preparation, women see these services and the proximity between transportation places and consumption places as an opportunity to optimize their time (Baffi 2012).

As this link between transportation and trade suggests, far from simply being places of circulation and movement, transportation places are also places of encounters, exchanges and interactions. The social function of the station was precisely analyzed by Jean-Fabien Steck (2004) through the Yamoussoukro bus station, a multiscale melting pot that represents as much a transportation place as a destination in itself for Ivorians. Jean-Fabien Steck also highlights the reflexive scope of the station, truly mirroring the transformations and tensions at work in society, which brings it closer to a public space. However, this analogy questions the very notion of public space, which does not always seem appropriate in understanding the modalities of urbanization in cities of Sub-Saharan Africa (Piermay 2002): in some cases, the dichotomy between public space and private space does not exist, and in other cases, the absence of public space testifies to the legacy of colonial powers that imposed their political domination through urban planning. This was the case in South Africa, where for decades apartheid urbanism was designed to minimize spaces of interaction of the segregated population in order to avoid the slightest desire for political gathering and ultimately uprising (Houssay-Holzschuch and Teppo 2009). As such, for several decades transportation spaces proved to be one of the only places for meeting and a certain social diversity in South African metropolitan areas (Baffi 2016). Thus, it was in public transportation that the social relations of segregated populations crystallized, giving rise to religious ceremonies, demonstrations against the regime, but also clashes between African, "Colored"[3] and Asian populations. During apartheid, these places offered a broader picture of the tensions at work in a South African society whose identity was denied by the State and apartheid planning. The transportation place was thus a place of expression, a social and political outlet and one of the constituent elements of an African culture burgeoning at that time (Freund 2007).

8.2.3. *Polarities within fragmented urban spaces*

As points of departure or arrival, transportation places stimulate mobilities and convey the flows that animate the city and the pace of urban dynamics[4]. Through them, centralities emerge, as well as spaces of relegation and new urbanization fronts. At city scale, the frequentation of stations and transportation places reflects the "spatial mismatch" that exists in many cities in sub-Saharan Africa (Naudé 2008; Lall et al. 2017), that is, the gap between job distribution and the distribution of places of residence. In Cape Town or Dakar and Abidjan, the stations in the outer suburban neighborhoods of the city constitute the main hubs emitting flow in the morning and the stations of the city center the receiving hubs. Moreover, the intensity of these pendulum movements is also observed through the distribution between peak and off-peak hours. Peak hours generally begin at dawn, and for several hours the stations are saturated, likewise in the late afternoon until early evening. These times of day when transportation places are overcrowded, during which operators take few breaks if any, are followed by times when stations become relaxing places for drivers, who spend time playing, sleeping or talking (Figure 8.1). The strong fluctuations between off-peak and peak hours even lead in some cases to

3 The "Coloreds" are descendants of the Malays, brought over as of the 17th century by Dutch settlers, and who gradually mixed with the South African population. "Coloreds" historically have a strong presence in the city of Cape Town.

4 In cities where there are sometimes few statistics and where the data are often fragmented, it is also frequently at the scale of stations and transportation places that counts are carried out in order to understand travel flows.

the disappearance of stations once the peak hour is over, as in Abidjan (Kassi-Djodjo 2015).

Figure 8.1. *Drivers play cards during off-peak hours at Station Deck minibus station in Cape Town (source: Baffi (2010)). For a color version of this figure, see www.iste.co.uk/fretigny/transportation.zip*

Lastly, the diffusion of transportation places at the scale of metropolitan areas provides information on the dynamics of urbanization and the emergence of new neighborhoods in line with intense urban growth. These new neighborhoods, some mainly composed of precarious habitats, are often located on the outskirts and have little or no connection to existing networks. These new urbanization fronts thus represent a new, major demand to be satisfied for paratransit operators. Lombard et al. (2004) illustrates this in Dakar: the establishment of new line terminals to developing neighborhoods, on the margins of the public transit service area, serves as a legitimation of their activity for paratransit operators. In Abidjan, Kassi-Djodjo (2015) points out that the correlation established between the emergence of new stations and the population growth of neighborhoods is tending to fade. The number of stations is now increasing primarily in neighborhoods where economic activities are concentrated at the expense of certain residential neighborhoods. In this scenario, it is therefore the very nature of the urbanization process that is illuminated through the paratransit service strategy. Kerzhner and Martens (2018) offer a similar analysis of the minibus sector in Lubumbashi: "Finally, the evolving, self-organizing nature of the layout of paratransit network in informal built environments raises questions as to the effects on land-uses and the development of urban commercial and industrial nodes and housing costs and densities. In this regard, drivers' decisions – such as when and where to start a new route or define a new terminal – may be playing a substantial role in the creation of urban form in fast growing cities". As such, paratransit services play a pilot role, as pioneer services preceding the creation of public transportation lines, the evolution of places connected reflecting the pace and dynamics of urbanization. Analyzing the metropolitan areas of the South through paratransit places offers, more generally, an understanding of the functioning of urban societies that can withstand the theoretical frames developed in cities and countries of the Global North.

8.3. Places with high stakes and power relationships

Whether it is a matter of making room for these places in the city, integrating them, or not, into planning, or deploying specific identity practices, the power games and the social and political dynamics that unfold in there prove to be intense.

8.3.1. *Places imposing their presence in the urban space*

Through their structural dimension, stations contribute to bringing visibility to the paratransit sector and its ecosystem, the interests of its actors and the flows that drive it. Yet, this affirmation of stations in the urban landscape is not without creating tensions: first because it is done at the expense of other urban actors; second because it raises the question of the adequacy of this sector with the standards and rules of place-making derived from international standards; and lastly because it highlights the dissensions that animate a complex and heterogeneous sector. Above all, it should be remembered that paratransit spaces are most often created spontaneously, from the regular use of vacant spaces in cities, whether in N'Djamena (Heinrichs et al. 2016), Dakar (Lombard et al. 2004) or Abidjan (Kassi-Djodjo 2015). While these spaces are gradually acquiring an official dimension, as is the case in Dakar or in the Dar es Salaam camps[5] in Tanzania, this is not the case in all cities. Moreover, the spontaneous emergence of these sites is often at the expense of other usages of public space and other individuals or groups of individuals. Conflicts of usage are all the more regular as transportation spaces are most often located near crossroads, intersections, or even in certain cases directly on roads. In N'Djamena, Heinrichs et al. (2016) specify that these conflicts of usage concern motorcycle taxis and traders, whose access to their stalls is hampered by parked vehicles. In this case, there is no obvious way out of the conflict, since

5 Spaces in which motorcycle-taxi drivers gather. Drivers must be co-opted to access these camps, which today constitute professional organizations recognized by the public authorities.

transportation operators and traders use public space for personal purposes, and local authorities then advocate a *laissez-faire* approach. These difficulties also concern the circulation of pedestrians in the vicinity of stations, complicated by a lack of signage which in some cases increases the number of accidents.

The informality in which these transportation places evolve is also the result of strategies by operators and professional organizations. Lombard et al. (2004) demonstrate that the absence of formalization benefits these actors, who thus avoid taxation. On the one hand, this strategy responds to the constraints of operators, who seek to optimize their economic model and minimize everyday expenses by avoiding taxes. On the other hand, this option is also an asset for professional organizations and trade unions, which are then able to establish a balance of power and enter into negotiations with public authorities to determine for themselves the amount of taxes to be paid in the absence of clear regulations. The legal and legislative vacuum in which stations develop moreover encourages the development of violent and even mafia practices. In Abidjan, Kassi-Djodjo (2015) mentions the role of actors, called *gnamboro*, who carry out veritable racketeering of transportation operators and local authorities for the benefit of trade unions. These individuals impose their control over stations using violence and have become essential actors in the sector. Seftel (2020) reports a similar functioning in South Africa, where professional organizations do not hesitate in some cases to resort to violence to establish their power and territorial control. Indeed, control of the "routes" (i.e. sections on which members of professional organizations are permitted to practice) is sometimes obtained by using the services of hired killers. But only the wealthiest organizations are able to use this type of service, thus increasing their hold and their financial resources.

This example, albeit extreme, reflects the intricacy of the rationales of power, criminality and informality that unfold in the "in-between" spaces created by paratransit. Transportation spaces are sometimes akin to lawless zones, in which informal and criminal activities may flourish. The porous borders that sometimes exist between paratransit activity and criminal activities are largely related to the significant flows of cash that transit on a daily basis and facilitate money laundering (Seftel 2020). More generally, the adjacency between these activities echoes the "in-betweenness" in which the sector evolves. This in-betweenness makes it possible to consider both the visible aspect of paratransit, that is to say the service provided, as well as the informality in which the sector evolves, and which questions the relationship to the norm. This relationship to the norm and to rules is particularly evident when paratransit activities and criminal activities contribute to creating a climate of insecurity that impacts urban safety in general (Kassi-Djodjo 2015).

8.3.2. *The shadow planning of paratransit places*

On a larger scale, beyond the analysis of their functioning and the actors who animate them, stations, stops or assembly points crystallize the interests and objectives of actors of the urban space. They are akin to objects of power and places of conquest. Indeed, the multiplying of these places and their diffusion across the urban space reflect the existence of territorialization processes and recall that if networks are "instruments of power" (Raffestin 1980), stations constitute the hubs of these networks, that is, their control points and their gateways. The control of stations and paratransit places thus becomes a strategic issue in the cities of the Global South. Through the control of these territories, it is the ability to impose standards on the service and the sector that is at stake. While the emergence of paratransit places denotes the economic and social weight of actors in this sector, the lack of consideration of these places, or their negative consideration, recalls in return the regulatory power of the various actors considered legitimate to decide what is formal and what is not.

While the unclear status of paratransit places benefits operators and professional organizations, it is also the subject of strategies on the part of regional and national public authorities as well as local authorities. In the case of Côte d'Ivoire, Kassi-Djodjo (2015) recalls that the current deployment of stations responds to the ban on paratransit services for several decades within the urban space to protect the activities of the public bus company, SOTRA. For a long time, *woro-woro* (collective taxis) and *gbakas* (collective minibuses) were only authorized to circulate in outer suburban spaces, until they were allowed access to the city center due to the decline of SOTRA and the lack of transportation supply. However, despite the evolution of regulations, stations and other transportation places do not appear on the city's development plans. The authorities' *laissez-faire* approach indicates tacit consideration of the role played by these transportation services in a context where public services are not able to meet travel demand, without fully recognizing the function of these services on the scale of urban space. Lastly, in some cases, the lack of formalization also reflects conflicts of scale or lack of coordination existing between the different territorial levels of the authorities themselves, as is observed in Dakar, where station management is regulated, but this can sometimes come under the responsibility of *arrondissement* (borough) municipalities, sometimes of cities (Lombard et al. 2004).

The lack of integration of paratransit services and actors into development plans and planning documents illustrates the clear lack of information on this sector (Klopp and Cavoli 2019). It also reflects the lack of consideration accorded to it by local decision-makers, in view of its low compliance with the standards and models of cities that they wish to develop. In some cases, paratransit territories are taken into account in urban development plans, as was the case in Cape Town. The renovation of a shopping mall located in the large African township of Khayelitsha resulted in the construction of a station for minibuses, with platforms and adapted signage. However, the planning for this place was not carried out in consultation with paratransit operators and it is located near a little-used entrance of the shopping mall, poorly connected to the township's street network. Thus, despite the existence of an equipped site (Figure 8.2), minibus operators continue to group together and load passengers from the entrance on the other side of the shopping mall through which many transit and are directed toward the densest neighborhoods of the township (Figure 8.3). In this case, the desire to plan the minibus station was hampered by a failure to consider the practices of users and the economic logics of paratransit actors. The result is an ambiguous situation where the informal minibus station is tolerated by the public authorities, while the infrastructure built on the other side of the shopping mall is appropriated by other groups of users, namely, shop customers and vendors, who park their vehicles there.

Figure 8.2. *Renovated minibus station at Ekuphumleni shopping mall in Khayelitsha, Cape Town (source: Baffi (2010)). For a color version of this figure, see www.iste.co.uk/fretigny/transportation.zip*

Lastly, it is also important to mention that in other shopping malls, minibus access is not tolerated (Figure 8.4). In this case, it is the negative externalities of the paratransit sector that are targeted (noise, pollution and possible congestion), but also the image of this mode of transportation, which does not conform to the standards and model of a modern, welcoming and clean shopping mall. Where local authorities sometimes struggle to impose their rules, economic actors do not hesitate to build enclaves sealed off from services, territorialization rationale and standards of the paratransit sector.

Figure 8.3. *On the other side of the shopping mall in Ekuphumleni: the minibus station occupied by operators (source: Ninot (2010)). For a color version of this figure, see www.iste.co.uk/fretigny/transportation.zip*

Figure 8.4. *Near Bayside shopping mall (Cape Town), access restrictions for minibus-taxis (source: Baffi (2013)). For a color version of this figure, see www.iste.co.uk/fretigny/transportation.zip*

8.3.3. *The sense of place of paratransit: a world in its own right*

The insufficiency, as well as the refusal, of the paratransit sector and its places with its prevailing rules and standards is a reminder of the informality to which this sector is assigned. For Steck (2015, p. 100), rather than understanding this informality in the light of the dichotomy it supposedly forms with the norms, we must focus on the gap that exists between the two and the way it "manages itself". In the case of the paratransit sector, this gap results in practices and standards that are specific to that sector and are observed at the scale of stations, transportation places and transportation spaces (which may be fixed or mobile, e.g., onboard vehicles), and in urban space, to the point of building a real sense of place related to the paratransit sector.

First of all, the functioning of the sector is organized based on what Zegras et al. (2015) call hidden information, which is shared by insiders. This hidden information relates to the service and its operation (destination of vehicles, organization of service on demand, rates applied, etc.). This is akin to vernacular knowledge difficult to decipher for the uninitiated. Inside the vehicle, veritable rituals govern the organization of the crew (the driver and the conductor) and passengers, whether it is which seat to sit on, depending on the destination of each one, the method of collecting cash between the passengers and the conductor, or the announcement of the desired stop. The reference points used by both crew and passengers correspond to specific objects and landmarks within the urban landscape, such as trees, specific buildings or traffic lights. The gestural semiology used by passengers and crew to communicate between the street space and that of the vehicle comes under the same logic. Woolf and Joubert (2013) study some of these gestures, used in Johannesburg. They highlight, on the one hand, the evolving aspect of this semiology, which becomes enriched as new neighborhoods are created, and its multicultural nature, on the other hand, since it is used by individuals who do not always speak the same language.

Next, paratransit vehicles themselves constitute places of identity anchorage within the urban space. In Senegal, for example, *cars rapides* are part of the visual identity of the city of Dakar, where vehicles are painted in specific bright colors and bear numerous inscriptions. Music played at high volume is also characteristic of this sector, as is the style of music chosen by drivers. More generally, many drivers and owners personalize their vehicles to display a cultural affiliation, to a football team, for example, or to a social group through the use of a specific iconography. Another aim of personalization is to place vehicles and their passengers under divine or mystical protection, whether by God, marabouts or other forms of belief. In Dakar, the omnipresence of certain inscriptions reminds us of the central importance of marabout brotherhoods in the paratransit sector, in particular for *cars rapides* (Godard and Teurnier 1992). The visual identity of the sector, the practices and rituals associated with it, like the specific modes of communication created, then contribute to the inclusion of paratransit in the urban space, but also to the construction of an original space of its own.

Lastly, taking an interest in paratransit places implies considering their multiscale anchoring, linked to the cultural, ethnic and identity dimension of the social groups that animate them.

This anchoring is revealed, for example, through the rationale animating the minibus wars in South Africa, from 1987 to 1994. Also known as the "taxi wars", this term refers to clashes shaking the townships of major cities and claiming thousands of casualties, notably among civilians. For several years, the conflict crystallized around route control: at the end of apartheid, minibuses were now authorized to serve territories located outside townships. This opening up of the market leads to violent clashes between operators, in dispute over the distribution of routes, especially for the most lucrative ones. But these clashes must be viewed from the perspective of the ethnic and political tensions across the country at that time, making paratransit services an issue and a tool for territorialization. For example, the clashes oppose supporters of the African National Congress and those of the Inkatha Freedom Party[6] in KwaZulu-Natal province. They also see different ethnic groups confronting each other, as in Cape Town, where the historical association of the local township opposes that of migrants from the Bantustans (Dugard 2001). The territory of the minibus associations therefore corresponds to the routes and perimeter that the operators are authorized to serve – according to the permits issued by the local authorities, but above all by virtue of the power relations between actors – and it also reflects the identity affiliation claimed by members of the associations. Far from being solely an economic conquest, the taxi war also denotes the political, ethnic and cultural dimension playing out in the background and the multiscale scope of the territorialization process. The sense of paratransit places therefore recalls a political space that goes beyond just the limits of the city. The affirmation of the paratransit sector and its places is thus particularly revealing of the relationships between actors at work in policies and practices. It also shows the hybridization processes operating within metropolitan spaces, and particularly in cities of the Global South.

---

6 The Inkatha Freedom Party is a Zulu political party in South Africa that advocates for territorial separatism for the Zulu population. Although initially close to the ANC, strong dissensions appeared at the turn of the 1990s, provoking a veritable civil war in the province of KwaZulu-Natal.

8.4. The evolution of paratransit, a mirror of urban space hybridization

This hybridization can be grasped in three ways: by urban production that proceeds by "institutional bricolage", by the implementation of hybrid transportation services combining paratransit and other transportation modes and, in recent years, by the role of digital tools, which renew the in-between function of paratransit places.

8.4.1. *"Institutional bricolage" as a mode of production of urban space*

In recent decades, paratransit services have come to be prevalent in most cities of the Global South, notably in Sub-Saharan Africa. Numerous regulatory measures have been introduced by local authorities, the aim of which is both to mitigate access to the sector and to reshape its functioning to mitigate its negative externalities, and in some cases to eliminate it. Many factors explain why these measures have not always been applied or have led to resistance, particularly when they were introduced too abruptly and without consultation with operators, as was the case in Dar es Salaam (Rizzo 2014), resulting in the failure of the Bus Rapid Transit[7] project in its first iteration.

One of the main obstacles to the introduction of these regulatory measures concerns the informality in which the paratransit sector evolves. Indeed, the organization and functioning of the sector are shaped by both formal and informal rules and codes. The paratransit sector catalyzes the vested interests of a variety of actors, from operators themselves to vehicle manufacturers and insurers, to civil servants and police in cities where corruption is widespread. This entanglement of practices and interests is evident in paratransit places, which constitute the nerve centers of the sector. From these places, it is possible to observe the superimposition of standards and rules as well as the implementation of arrangements, which lead to reference to "institutional bricolage" to evoke the established modes of regulation and to overcome the dichotomy between formal and informal (Cleaver 2001).

The notion of institutional bricolage allows consideration of the hybridization that occurs between the rules established through practice and by the operators

7 Bus Rapid Transport is a high-capacity (i.e. accommodating many passengers), own-site (i.e. with dedicated traffic lanes) transportation system. It differs from the "high level of service" bus (BHNS: *Bus à Haut Niveau de Service*) model in that it focuses primarily on transportation infrastructure in itself, often leaving aside the question of urban integration; as such the BHNS also features as a tool for transportation planning and urban development.

themselves, on the one hand, and the regulations drawn up by the public authorities, on the other hand.

Heinrichs et al. (2017) illustrate this based on the regulation processes for motorcycle-taxis in Dar es Salaam. Motorcycle-taxis gather in camps, that is, spaces created spontaneously, where drivers park when they are not undertaking rides. Camp members are co-opted and obey a number of rules, with respect to routes and fares, for example. They benefit from a form of protection provided by the group, particularly in the event of conflicts with users. Many of these camps, some of which benefit from "association" status, with a chairman, treasurer and secretary, also register with the local authorities, the *mtaa*. The recognition and formalization of these camps in turn shape the criteria for obtaining an operating licenses, as set by the local authorities. The motorcycle-taxi driver must specify which camp they are affiliated to (Heinrichs et al. 2017). Local authorities thus rely on mechanisms established informally to regulate the sector, effectively formalizing bottom-up processes. The fact remains that in Dar es Salaam, this regulation of motorcycle taxis excludes many operators who do not belong to a camp. In other cases, formalization and hybridization operate according to top-down processes, notably through public authorities equipping paratransit places with brick-and-mortar installations, toilets, signage or solar panels. By recognizing the existence and function of the place, the objective of the public authorities is to structure and formalize the use of the equipment and its maintenance.

The above-mentioned example of Khayelitsha shopping mall revealed the inability of the public authorities to take into account the rationale of users and operators in the planning of the transportation place. But for several years, new stations have been built or renovated in Cape Town, some of which are equipped with solar panels and rainwater recycling devices, such as in the townships of Wallacedene (SA Roofing 2014) and Nomzamo (City of Cape Town 2015), testament to a formalization process on the part of local authorities. These developments contribute not only to the recognition of the sector, but also to a better linking of transportation places with the surrounding spaces and with the practices of city dwellers, for whom these stations become points of access to resources. Lastly, through the notion of institutional bricolage, a key characteristic of urbanization in cities of the Global South transpires: the existence of multiple powers – notably of private actors – whose negotiations, arrangements and resistance shape urban space (Jacquot et al. 2016).

8.4.2. *Formalizing informality: planning hybrid services*

Institutional bricolage does not just relate to the practices and rules governing stations. It also refers in some cases to tacit recognition of the services provided by paratransit operators, as in Cape Town. Since 2011, paratransit vehicles (minibuses) are no longer permitted to travel on certain sections of the road since the introduction of a "Bus Rapid Transit (BRT)" in order to limit competition. However, after several years, the BRT service has proved insufficient in size on one of the main sections of the metropolitan area's roads during rush hour. Moreover, the higher fares across these time periods, established to encourage a better distribution of demand throughout the day, prove prohibitive for some households. Thus, the municipality tolerates the reappearance of minibuses on this section of road, where operators are tacitly authorized to load waiting passengers[8].

While this recognition of the usefulness of the paratransit service is mostly tacit, the municipality of Cape Town is now considering establishing a "hybrid" service. This reflection is based on the observation that, despite the introduction of the BRT service, minibuses remain an essential component of the mobility system in Cape Town and will not disappear (Lesteven and Boutueil 2017). On the contrary, after several years in operation, the limits observed with the BRT service attest to this mode of transportation unsuitability with respect to certain characteristics of South Africa's cities, in any case as the sole mode of public transportation at the scale of the metropolitan area (Scorcia and Munoz-Raskin 2019). Thus, the concept of a "hybrid" service proposes adjusting the complementarity of the modes: in this configuration, capacity transportation serves the main sections of the urban space's roads, and minibus services provide the feeder services, or supplement the supply of capacity transportation during rush hour (Salazar-Ferro and Behrens 2015). For the time being, such hybrid services largely remain to be built in Cape Town. They are already planned in other cities such as Dakar, where the newly built networks of regional and local rail routes and BRT will rely on feeder services operated by paratransit operators.

Furthermore, it is significant that this "hybrid" model is based on characteristics of paratransit services such as the adaptability and flexibility of the service, while evolving the economic model and establishing fixed timetables and routes for operators. This standardization process ultimately leads to the retention of certain characteristics of paratransit services at the expense of others, and thus helps to favor certain actors, transportation places and territories over others. This is what Morange suggests in her definition of emergence (2011, p. 349). She returns to the etymology of this term to designate the mechanisms that lead certain territories to

8 Interview with a representative of the Municipality of Cape Town, conducted in 2018.

"emerge" from underdevelopment, and thus insists on the differentiation mechanisms that exist between territories. In the case of paratransit services, the emergence of these "hybrid" transportation services could then come at the expense of services still immersed in informality (*amaphelas*, for example), reinforcing the mechanisms of differentiation between actors and services and discontinuities in the urban space. While the paratransit sector is tending to gradually become normalized by adapting to standards that seek to reduce informality, in return, the integration of this sector contributes to redefining, transforming and hybridizing the modes of production of urban space.

8.4.3. *The blurring effect of digital tools*

In this hybridization of urban services, digital tools have played a central role for several years, as reflected in the formula of "digital revolution", used to evoke the transformations at work in the cities of the Global North and South (Peyroux and Ninot 2019). More specifically, in the urban transportation sector, the combination of mobile tools and the use of geolocation open up new prospects. In the mobility sector, for example, it is now possible to carry out structured surveys on paratransit vehicles based on the tracking data from GPS, thus opening up possibilities for operators to optimize vehicle operations, improved transportation regulation for public authorities and better knowledge of daily revenue for vehicle owners. But the introduction of these tools also shifts the balance of power between actors in the sector, especially when it catalyzes numerous interests and financial flows. Thus, drivers are not always inclined to share the amount of revenue generated daily, especially with vehicle owners (at the risk of seeing the share to be paid back to the owner increase). More generally, the introduction of these tools tends to modify the relationship between actors and modes of operation within the sector, potentially calling into question the role of transportation places.

Indeed, in addition to the transformation of governance and interactions between actors, we also observe an evolution of the supply and working conditions of operators. The few barriers to entry into the urban mobility sector explains the explosion in the number of operators in recent years, supported by numerous platforms such as SafeBoda, Heetch, and Gozem, for motorcycle taxis; Uber, Bolt and Yango for taxis; and SWVL for on demand buses. Yet, the arrival of new services often occurs at the expense of operators, in a context where internal competition is intense, while the regulatory methods introduced by platforms tend to reinforce the precariousness for operators (Rekhviashvili and Sgibnev 2020).

Beyond drivers' working conditions, the introduction of these digital tools calls into question the very existence of certain related professions, such as "coxers": in the application-driven system, the digitization of supply meeting demand renders obsolete their role, which consists of directing passengers and ensuring a good vehicle filling rate. More generally, through the use of these tools, the function of transportation places as regulatory spaces for the sector is also called into question. The deployment of these services often takes place outside the regulatory frameworks and practices established by professional organizations, in particular to obtain licenses and operational areas. In Cairo, the SWVL platform sparked controversy by basing its development strategy on obtaining licenses granted to tourist vehicles, only to in reality use them much more widely to benefit those living in the city (Schalekamp and Saddier 2020).

Lastly, the practices induced by the use of digital tools also lead to the redefinition of transport spaces in urban areas. Thus, a corollary of the controversies caused by the SWVL application in Cairo concerned parking problems for the platform's vehicles, for which responsibility then fell to the driver with no dedicated space or parking lot provided by the company. The absence of stopping points for e-hailing drivers is reminiscent of the practices of the *amaphela* drivers in South African townships, who crisscross the roads in search of users. In some cases, however, these digital services are materialized in the urban space. At Cape Town airport, for example, a service area reserved for e-hailing[9] has been in place since 2017. It indicates the pioneering nature of cities of the Global South, where on-demand transportation has existed for a long time (Figure 8.5).

If these platforms form part of an international turning point and assume the guise of "modernity" (Tanikawa et al. 2022), they also renew practices involving institutional bricolage while freeing themselves of territorial frameworks (stations, camps and other anchor points, as well as the territories served), contributing to transforming the regulatory processes and modes of production of urban services.

Between the adoption of international trends and the readapting of local innovations, digital tools thus contribute to blurring development trajectories. The in-betweenness of paratransit places, between material and virtual space, thus reflects the process of economic and social transition in which Sub-Saharan African cities find themselves.

9 That is, hailing a vehicle electronically.

Figure 8.5. *A space dedicated to the pick-up of e-hailing passengers at Cape Town Airport (source: Baffi (2018)). For a color version of this figure, see www.iste.co.uk/fretigny/transportation.zip*

8.5. Conclusion

Far from being "non-places" (Augé 1992), the mobility turn has proposed, since the early 2000s, considering transportation spaces not only as vectors of movement, but also as anchoring points whose materiality depends on the interplay of actors (Sheller 2015), and which are territorial constructions in their own right (Jensen 2009). Indeed, on a global scale, transportation spaces provide a picture both of the predominant place held by communication networks in our societies, and of the configurations and specific local and regional appropriations. During the first months of the Covid-19 pandemic in 2020, images of deserted stations in European capitals contrasted with the queues to enter vehicles in many cities in Sub-Saharan Africa. However, even if the significance of these places on the scale of contemporary societies is no longer to be demonstrated, their analysis remains complex, especially in cities of the Global South, where paratransit services still often represents the main transportation offer.

Indeed, this complexity is due in particular to the multiplicity of functions, uses and actors that animate these places. By presenting themselves as "in-between", paratransit places thus assert themselves as fitting observatories for understanding certain modalities of urbanization in the Global South. In particular, the diversity of these places recalls the continuum of configurations that characterize paratransit services, from informal roadside stops to multi-modal hubs. Whatever their degree of formalization, these places very often play an all the more fundamental role, especially in the context of cities of the Global South. The observatory dimension of paratransit places is also revealed through the territorialization processes that animate these places, led by actors whose standards and interests are at times antagonistic, and through the appropriation of exogenous standards and technologies that play out there. The bricolage and hybridizations that come about as a result are observed today, for example, both in minibus stations in Southern African cities and in gatherings of motorcycle-taxi drivers in Western Africa, which are all manifestation of societies and territories in transition.

Thus, in these cities perhaps more than elsewhere, transportation spaces offer a picture of certain early signs and underlying trends of territorial transformations at work. Their diversity and complexity remind us, lastly, of the need to multiply the frames of reference and disciplinary frameworks (geography, urbanism, sociology, anthropology, economics, etc.) to understand as finely as possible the development trajectories that are emerging.

8.6. References

Augé, M. (1992). *Non-lieux. Introduction à une anthropologie de la surmodernité*. Le Seuil, Paris.

Baffi, S. (2012). Pratique d'une métropole émergente par les usagers des transports en commun, le cas du Cap, Afrique du Sud. *EchoGéo*, 21. doi: 10.4000/echogeo.13139.

Baffi, S. (2016). Le chemin de fer et la ville dans les processus de territorialisation en Afrique du Sud : de la séparation à l'intégration territoriale ? PhD Thesis, Université Paris 1 Panthéon-Sorbonne, Paris.

Baffi, S. (2017). De la mobilité à l'accessibilité : l'évolution des transports comme ressources dans la ville sud-africaine. *EchoGéo*, 40. doi: 10.4000/echogeo.14931.

Bertrand, M. (2010). Introduction. *Revue Tiers-Monde*, 201(1), 7–23.

Chaléard, J.-L. (ed.) (2014). *Métropoles aux Suds, le défi des périphéries ?* Karthala, Paris.

City of Cape Town (2015). City opens green taxi rank with solar panels, kiosks and washing facilities in Nomzamo [Online]. Available at: https://www.tct.gov.za/en/news-and-alerts/press-releases/articles/city-opens-green-taxi-rank-with-solar-panels-kiosks-and-washing-facilities-in-Nomzamo/article-copy/ [Accessed 20 July 2021].

Cleaver, F. (2001). Institutional bricolage, conflict and cooperation in Usangu, Tanzania. *IDS Bulletin*, 32(4), 26–35.

Delaunay, T. and Baffi, S. (2020). Les impacts de la crise sanitaire du Covid-19 sur les systèmes de mobilité en Afrique : dépendance aux transports collectifs, résilience des artisans, mutations de la gouvernance et diffusion des innovations [Online]. Available at: http://www.codatu.org/actualites/les-impacts-de-la-crise-sanitaire-du-covid-19-sur-les-systemes-de-mobilite-en-afrique-dependance-aux-transports-collectifs-resilience-des-artisans-mutations-de-la-gouvernance-et-diffusion-des-inn/ [Accessed 16 July 2021].

Dugard, J. (2001). From low intensity war to mafia war: Taxi violence in South Africa (1987–2000). Report, Centre for the Study of Violence and Reconciliation, Cape Town [Online]. Available at: https://help.csvr.org.za/wp-content/uploads/2001/01/fromlowintensity.pdf [Accessed 16 July 2021].

Farrington, J.H. (2007). The new narrative of accessibility: Its potential contribution to discourse in (transport) geography. *Journal of Transport Geography*, 15, 319–330.

Fol, S. and Gallez, C. (2013). Mobilité, accessibilité et équité : pour un renouvellement de l'analyse des inégalités sociales d'accès à la ville. In *Colloque Enjeux interdisciplinaires émergents pour comprendre, projeter et fabriquer la ville de demain.* LabEx Futurs Urbains, Marne-la-Vallée [Online]. Available at: https://halshs.archives-ouvertes.fr/halshs-00780292/.

Freund, B. (2007). *The African City: A History.* Cambridge University Press, New York.

Godard, X. (1987). Transports privés, transports publics, expériences de villes du Tiers Monde. INRETS. *Synthèse INRETS No. 3*, Bron.

Godard, X. and Teurnier, P. (1992). *Les transports urbains en Afrique à l'heure de l'ajustement : redéfinir le service public*. Karthala, Paris.

Guézéré, A. (2012). Territoires des taxis-motos à Lomé : de la pratique quotidienne à la recomposition des espaces urbains et des liens sociaux. *Géographie, économie, société*, 1(1), 53–72.

Heinrichs, D., Goletz, M., Lenz, B. (2016). Negotiating territory: Strategies of informal transport operators to access public space in urban Africa and Latin America. *Transportation Research Procedia*, 25, 4507–4517.

Heinrichs, D., Ehebrecht, D., Lenz, B. (2017). Moving beyond informality? Theory and reality of public transport in urban Africa. In *Urban Mobilities in the Global South*, Uteng, T.P. and Lucas, K. (eds). Routledge, London.

Houssay-Holzschuch, M. (2006). *Antimondes. Espaces en marges, espaces invisibles.* L'Harmattan, Paris.

Houssay-Holzschuch, M. and Teppo, A. (2009). A mall for all? Race and public space in post-apartheid Cape Town. *Cultural Geographies*, 16(3), 351–379.

Jacquot, S., Sierra, A., Tadié, J. (2016). Informalité politique, pouvoirs et envers des espaces urbains. *L'Espace Politique*, 29. doi: 10.4000/espacepolitique.3805.

Jensen, O.B. (2009). Flows of meaning, cultures of movements–urban mobility as meaningful everyday life practice. *Mobilities*, 4(1), 139–158.

Kassi-Djodjo, I. (2015). Foisonnement des gares spontanées et recomposition du paysage urbain d'Abidjan. *Le journal des Sciences sociales*, 12(2), 165–176.

Kerzhner, T. and Martens, K. (2018). The role of labour informality in paratransit services: Case study of Lubumbashi, Democratic Republic of Congo. *Transportation Research Board Annual Meeting*. Washington, D.C.

Klopp, J. and Cavoli, C. (2019) Mapping minibuses in Maputo and Nairobi: Engaging para-transit in transportation planning in African cities. *Transport Reviews*, 39(5), 657–676.

Lall, S.V., Henderson, V.M. Jr, Venables, A.J. (2017). Africa's cities: Opening doors to the world. Report, Banque mondiale, Washington, D.C.

Lammoglia, A. (2013). Analyse et modélisation multi-agents de transports flexibles. Comparaison de services français et sénégalais. PhD Thesis, Université d'Avignon, Avignon.

Lesteven, G. and Boutueil, V. (2018). Is paratransit a key asset for a sustainable urban mobility system? Insights from three African cities. *Transportation Research Board Annual Meeting*, Washington, D.C.

Lombard, J. and Ninot, O. (2012). Des mobilités aux transports. Regards croisés en Afrique de l'Ouest. *Echogéo*, 20. doi: 10.4000/echogeo.13127.

Lombard, J., Sakho, P., Bruez, F., Baillon, A. (2004). Les rues de Dakar : espaces de négociation des transports collectifs. *Autrepart*, 4(4), 115–134.

Lomme, R. (2008). Should South African minibus taxis be scrapped? Formalizing informal urban transport in a developing country. In *CODATU XIII Conference*, Ho Chi Minh City.

Monnet, J. (2007). Ambulantage et métropolisation. *Cybergeo*. doi: 10.4000/cybergeo.5565.

Monnet, J. (2010). Le territoire réticulaire. *Anthropos*, 227, 91–104.

Morange, M. (2011). "Émergence locale" et régénération urbaine au centre-ville du Cap. *Bulletin de l'Association de Géographes Français*, 88(3), 348–360.

Naudé, W. (2008). Is there a spatial mismatch in South Africa's metropolitan labour market? *Cities*, 25, 268–276.

Paquette, C. (2010). Mobilité quotidienne et accès à la ville des ménages périurbains dans l'agglomération de Mexico. Une lecture des liens entre pauvreté et mobilité. *Revue Tiers-Monde*, 1, 157–175.

Peyroux, É. and Ninot, O. (2019). De la "smart city" au numérique généralisé : la géographie urbaine au défi du tournant numérique. *L'information géographique*, 83(2), 40–57.

Piermay, J.L. (2002). L'invention de la ville en Afrique subsaharienne. *Historiens et géographes*, 379, 59–65.

Raffestin, C. (1980). *Pour une géographie du pouvoir*. Lexis, Paris.

Rekhviashvili, L. and Sgibnev, W. (2020). Theorising informality and social embeddedness for the study of informal transport. Lessons from the marshrutka mobility phenomenon. *Journal of Transport Geography*, 88. doi: 10.1016/j.jtrangeo.2019.01.006.

Rink, B. (2022). Capturing amaphela: Negotiating township politics through shared mobility. *Geoforum*, 136, 232–241. doi: 10.1016/j.geoforum.2020.06.010.

Rizzo, M. (2014). The political economy of an urban megaproject: The bus rapid transit project in Tanzania. *African Affairs*, 114(455), 249–270.

Roy, A. (2005). Urban informality. Toward an epistemology of planning. *Journal of the American Planning Association*, 71(2), 147–158.

SA Roofing (2014). First "green" transport facility in South Africa unveiled [Online]. Available at: https://www.victronenergy.com/upload/documents/20141010_SAR_OCTOBER_2014_PG-6-9_TAXI_RANK.pdf [Accessed 20 July 2021].

Salazar-Ferro, P. (2015). Le transport collectif artisanal : une composante essentielle dans un système dual. Report, CODATU, AFD, Paris [Online]. Available at: http://www.codatu.org/bibliotheque/doc/le-transport-collectif-artisanal-une-composante-essentielle-dans-un-systeme-dual/ [Accessed 16 July 2021].

Salazar-Ferro, P. and Behrens, R. (2015). From direct to trunk-and-feeder public transport services in the Urban South: Territorial implications. *Journal of Transport and Land Use*, 8(1), 123–136.

Schalekamp, H. and Saddier, S. (2020). Emerging business models and service options in the shared transport sector in African cities. Report, Volvo Research and Educational Foundations, Göteborg.

Scorcia, H. and Munoz-Raskin, R. (2019). Why South African cities are different? Comparing Johannesburg's Rea Vaya bus rapid transit system with its Latin American siblings. *Case Studies on Transport Policy*, 7(2), 395–403.

Seck, S.M. (2006). Transport et territoires : les "coxeurs" de Dakar, du bénévolat à la prestation de service. *Bulletin de la Société Géographique de Liège*, 48, 7–17.

Seftel, L. (2020). Do windows for change exist in the minibus taxi industry? In *Inaugural Workshop of the African Urban Mobility: Past, Present and Future, Mobility and Access in African Cities Network*. University of the Witwatersrand, Johannesburg.

Sheller, M. (2015). Mobilizing transportation: Transporting mobilities. In *Transport, Mobility, and the Production of Urban Space*, Prytherch, D. and Cidell, J. (eds). Routledge, London.

Steck, J.-F. (2004). Requiem pour une gare routière : succès et fragilités d'un territoire informel à Yamoussoukro. *Autrepart*, 4(4), 95–114.

Steck, J.-F. (2015). Les villes informelles. Développement, aménagement, débats. HDR Thesis, Université Paris Ouest Nanterre – La Défense, Nanterre.

Tanikawa, K., Coulaud, L., Ninot, O. (forthcoming). *Révolution des transports et transition numérique dans les villes du Sud : vers un modèle de mobilité intelligente ?* ISTE Editions, London.

Wester, L. (2018). Transports collectifs et initiative individuelle. Approche des transports collectifs artisanaux par l'auto-organisation et les systèmes multi-agents. PhD Thesis, Université Aix-Marseille, Aix-en-Provence.

Woolf, S.E. and Joubert, J.W. (2013). A people-centred view on paratransit in South Africa. *Cities*, 35, 284–293.

Zegras, C., Eros, E., Mehndiratta, S. (2015). Digitalised transportation data as disruptive innovation: Prospects in Mexico City. Working Paper, Massachusetts Institute of Technology.

9

From Rapprochement to Separation: Transportation Places and Mobile Segregation

Jean-Baptiste FRÉTIGNY
PLACES, CY Cergy Paris Université, Cergy-Pontoise, France

9.1. Introduction

While transportation connects, it also participates in inequalities and as such contributes to the separation of individuals. Inequalities in transportation and mobilities are often discussed from the perspective of uneven accessibility to places. Residential areas are served very differently by transportation networks, making access to key resource places for work, training, socializing or sport and cultural activities more or less challenging. This approach to inequalities, rich in planning and inclusion issues, proves essential (Fol and Gallez 2017).

To fully appreciate the strength of segregation in mobility, this chapter develops another, complementary, approach. Segregation can be defined as the distancing of individuals and groups along lines of class, race, gender, disability, nationality, age, sexuality, etc. It is mainly considered from a residential point of view, pointing to the contrasting geographies of residence of different social groups. This framing has contributed to neglecting less rigid forms of segregation, linked to the mobility of individuals.

In this chapter, I adopt the perspective of a mobile segregation[1] in order to spotlight how transportation places act as material and symbolic operators of social stratification of societies, playing a full role in shaping mobilities and asserting social order. Through these places, many existing forms of segregation are naturalized and depoliticized, in the sense that they are exercised as if they were self-evident, and thus more easily reproduced. These places are not just an echo chamber and the segregation processes that unfold there reveal the complexity of power geometries. Studies carried out in a wide variety of directions, often without explicitly formulating such a research agenda, contribute to underpinning this investigation of the power dynamics at work in and through mobility.

Due to its heterogeneity and the large number of populations practicing it, transportation is at the center of multiple travel services but also differentiations, including market segmentations, whose scope is not solely operational. These differences are indeed part of a broader symbolic economy, that is, systems of production and consumption of public or private places and services that engage the identities, representations and hierarchies of societies (Zukin 1995; Harvey 2001). While the notion of symbolic economy has often been mobilized to grasp the role of economic sectors such as art, entertainment, fashion, tourism or real estate, in relationship with gentrification and urban changes, it also allows a different reading of the economic logics associated with transportation and mobilities, traditionally examined through the prism of socioeconomical analysis of transportation[2].

The investigation developed in this chapter enriches discussions on mobility and transportation justice (Sheller 2018; Verlinghieri and Schwanen 2020; Gallez 2023), drawing inspiration from political philosophy and feminist studies. It joins the discussions on social justice (Fraser 1996; Fainstein 2004), which focus on the fairness of resource distribution and, more broadly, of the social positions occupied by the members of a society. It thus substantially addresses the issues of individuals' agency, the recognition of their diverse identities, situations and practices, as well as the equitable distribution of the (dis)advantages of all kinds associated with mobilities. As concerns about climate, energy crises, resource depletion and biodiversity loss intensify, inequalities in (im)mobility[3] are also linked to environmental injustices.

1 The Mobiliscope (2023), a tool for tracking hourly variations in the population present during throughout the day, neighborhood by neighborhood, using data from mobility surveys, is another illustration of a renewing of segregation approaches through mobilities.

2 Originating from traffic engineering, socio-economics of transportation aims to model travel to guide planning and management projects for transportation infrastructure.

3 The prefix "im" is in parentheses (as per the choice of the publisher) but the use of the term "(im)mobility" in the book aims to evoke both immobility and mobility, granting them equal importance.

Mobility and transportation should also be seen as processes and practices of energy transformation, involving variable volumes of fossil resource extraction and combustion, with profoundly unequal repercussions. From an environmental perspective as well, transportation places can separate individuals.

To shed light on the separations operated by transportation places, we will first examine the social contrasts that play out between different transportation places before focusing on the vigor of segregations within each transportation place. The conclusion will finally allow to further nuance the analysis of these distancing processes.

9.2. Disjunctions in the frequentation of transportation places

A first approach to understanding segregation at work in and through transportation places consists of reflecting on their contrasting frequentation by different mobile publics, notably based on their social position. These disjunctions are part of what Graham and Marvin (2001) named "splintering urbanism", linked to unequal services provided by technical networks, among which transportation networks feature prominently.

9.2.1. *The social marking of transportation places*

The different modes of transportation and associated places, static or mobile, contribute to contrasting social worlds, in the sense that their frequentation is associated with publics whose profiles are not homogeneous across modes of transportation. Our thoughts readily turn to the "architectural apartheid" faced by the disabled (Imrie 2001), the development of transportation places frequently excluding them, due to the absence of elevators, escalators, visual and tactile cues, etc. Social class segregation is also very prevalent. Despite the recurring rhetoric on the "democratization of air transport", presented as a completed process, the air passenger population is far from reflecting the world population as a whole. The "we" of the advertisement in Figure 9.1, which moreover is associated with the airline's business-class cabin, is in reality aimed at a limited fraction of the planet's multiple social worlds. The analysis of the 2011 national survey of air passengers conducted by the French civil aviation authority among passengers aged 15 and over shows that, among passengers departing from Paris-Charles-de-Gaulle airport residing in mainland France, executives, senior intellectual professions and business leaders account for 40.2% of flights taken in the year preceding the survey, more than four times their proportion in the resident population aged 15 and over (9.7%). In contrast, blue-collar workers account for only 0.8% of the total, while they represent 12.4% of the population (Frétigny 2013). The increase in air trips in

France has not led to an air-passenger diversification of the same magnitude, as Demoli and Subtil (2019) showed over the period 1974–2008. It is also estimated that 2–4% of the world's population took an international flight in 2018 and that nearly 1% of the world's population is responsible for 50% of CO_2 emissions from commercial aviation (Gössling and Humpe 2020). Unsurprisingly, the unequal recourse to air transportation has become a key argument of environmental activists, calling for greater regulation and a reduction of air travel as part of the fight against climate change (see, for example, Hopkinson and Cairns 2021).

Figure 9.1. *"We are all connected": Who does the "we" in this advertisement at Nuremberg airport address? (source: Frétigny (2022)). For a color version of this figure, see www.iste.co.uk/fretigny/transportation.zip*

Various analyses have pointed to the global exacerbation of inequalities, documenting the rise of transportation places aimed at the wealthiest, "super-rich" fractions of society such as "super-yachts"[4], luxury cruise ships (Spence 2017) and private jets, part of "secessionist mobility spaces" (Budd and Hubbard 2010, p. 100). More discreet places are also involved, such as free ports, which play a key role in the (im)mobility of their valuables. These sites allow owners to be exempted from customs duties and other taxes on artworks and other collectibles as long as they are stored there. This advantage was originally granted due to the transit status of these goods to favor their circulation and thus commercial activity. Significantly, free ports were then positioned in coastal areas. However, their usage has shifted from this initial objective to exploiting the immobility of stored items for fiscal purposes (Schwarzkopf and Backsell 2021). Such places of transportation deceleration are

4 With a length of more than 40 m and possibly reaching 180 m or even more.

thus concentrated in territories such as Switzerland, Luxembourg, New York, Delaware, Singapore and Hong Kong. In addition to storing objects, free ports also house galleries and offices of art dealers. They enable artworks to change hands without leaving the free port and even, in some jurisdictions, exempt transactions from VAT.

At the other end of the social hierarchy, practices of tax avoidance and exploitation affect goods-related transportation places. The widespread use of flags of convenience in international maritime shipping is a case in point. By flying the flag of a lowest-bidder country regarding labor law, taxation or ship safety, shipowners assign their fleet to a jurisdiction with minimal regulations. The union-based International Transport Workers' Federation regularly updates a list of these flags of convenience, which currently comprises 43 states (ITF 2024), including Panama, Liberia and the Marshall Islands. Guillot-Wright (2021) conducted an almost year-long ethnographic survey of Filipino sailors from a flag-of-convenience vessel returning weekly to a US port in the Gulf of Mexico. Life aboard the ship is marked by structural violence, with the risk of being injured by moving ropes or being sent overboard in rough weather (tropical storms in particular), aggravated by the sailors' lack of sleep, reducing their vigilance. Fearing being declared unfit in hospital and losing their jobs, sailors treat their injuries themselves, without anesthetic. The ship's captain, worried about not returning to port on schedule as every Monday, which would result in his dismissal, even suffered a heart attack, a stress-related pathology identified on a recurring basis by lawyers, sailors and captains interviewed. Despite being major vehicles of globalized flows, these international transportation places remain excluded from the United Nations-defined right to decent work. Transportation places, even at finer scales, can foster intense labor exploitation, as demonstrated by the logistics sector (see Chapter 4).

Transportation places, in all their diversity, more broadly refract social hierarchies, giving them varying levels of material and symbolic visibility. While coach stations occupy an important place in cities such as Mexico City or Yamoussoukro, their presence in London or Paris is far more discreet. Despite being London's largest coach station with a distinctive Art Deco architecture, Victoria station is very much in the background, in the urban landscape and in the dominant representations of the city, compared to the city's rail stations, starting with Victoria rail station. The lack of public attention paid to these sites is even more striking in Paris. The international terminal of Gallieni, established by the coach company Eurolines in an already hardly hospitable bus station designed for city buses, closed in 2020 after the carrier went bankrupt. Paris-Bercy bus station, which took over from Gallieni, is an equally basic underground terminal with minimal facilities. No waiting room, few benches, poor ventilation of the exhaust gases generated by coaches, etc., despite serving 2–2.5 million passengers (more than half of France's coach passengers!) and offering routes to 299 destinations, including 175 international ones, in 2021 (ART 2022).

These inequalities also permeate transportation places associated with shorter range mobilities. Car ownership and use, linked to the importance of the automobile system, remain marked by class relations. In Singapore, for example, to limit road congestion in the city-state, car imports are heavily taxed. This measure is combined with an auction system for car registrations, with the number of annual registrations being subject to a quota, creating scarcity and de facto ensuring that car ownership remains limited to the wealthiest. This strongly reinforces the distinctive nature of car use (Lin 2012). On Réunion Island, France's overseas territory in the Indian Ocean, despite the prominence of automobility, three in 10 adults do not have a driver's license, as well as 40% of job seekers (SMTR 2018), often limiting their ability to find employment. The contrast between motorists, on the one hand, and the "captive" customer base of public transportation – including the "yellow coaches" that circle the island – and pedestrians along the roads, on the other hand, is deeply ingrained in the island's collective representations. Only the most central lines in the capital, Saint-Denis, show more diverse social uses, especially the bus rapid transit line from the East stadium to city hall (Frétigny et al. 2024). Beyond these striking examples, it is clear that motorization remains the norm in many parts of the Global North. Even if it dramatically marks the centers of the largest metropolitan areas, de-motorization is at this stage proving globally limited (Emery et al. 2021).

These logics of social hierarchy can devalue public transportation or walking. Passalacqua (2011, p. 269) notes that at the height of their popularity in the 1920s, the tramways in Paris were not allowed to cross the Arch of Triumph square at its intersection with Avenue Foch, the widest street of the city. The avenue was considered too prestigious for this mode of transportation, deemed unattractive by decision-makers and associated with the working class, blue-collar workers and the suburbs. The same ban was imposed at other large Parisian squares, such as Place de la Bastille, Place de la Nation and Place de la Madeleine, which weakened the use of trams both practically and symbolically.

These hierarchies of transportation and their places are also deeply entangled with gender. In Vienna, Eva Kail and Jutta Kleedorfer, feminist engineers from the city planning department, organized a pioneering and highly influential exhibition in 1991 (Jackowska and Novas Ferradás 2022). It highlighted the gendered nature of transportation modes, attested by numerous studies across the world. At this time, male users made up 62% of motorists and 65% of cyclists, while public transportation and walking were predominantly practiced by females, at 59% and 64%, respectively. The exhibition also stressed that streets were allocated 72% to roadways and car parks, leaving only 28% of the space for sidewalks and pedestrians, raising the issue of women's place in the city, as they were, and are still, more likely to walk[5]. The exhibition also denounced urban planning priorities that favored male-dominated home-to-work commutes to the detriment of other mobilities, more often engaged in by women and involving nearby, on-foot activities. Women accounted for 72% of trips for shopping, 78% for taking children to school and 65% for trips to the park. These findings still relevant today illustrate the importance of explicitly considering gender (in)equality in the development of all public policies, otherwise known as an integrated approach to gender, or gender mainstreaming.

The example of class and gender shows how deeply transportation places are engaged in socially contrasting mobile cultures and networks of exchange. However, segregation processes are complex, as they involve not only separation but also interdependence between populations.

9.2.2. *Separation and interdependence: the ambivalence of roads*

Roads, given their central role in a no less important automobile system, illuminate these situations combining separation and interdependence, in two ways. The first concerns the consequences of constructing and using road infrastructure for neighboring communities. The erection of new highways can lead to significant social harm, such as the 27-km elevated toll expressway in Nairobi between Jomo-Kenyatta Airport, the business district and affluent residential neighborhoods of the capital, completed in 2022. Denounced as a "road for the rich", this infrastructure is financed by the state-owned China Road and Bridge Corporation, the toll recipient. It cuts through the self-built Mukuru Kwa Njenga district and in October and November 2021 its construction spawned the demolition of homes for nearly 40,000 residents, as well as numerous shops, with no prior notice or alternative accommodation proposal (Ram 2021). This example shows the severe dispossession dynamics triggered by large-scale road infrastructure projects.

5 Frequently used in the field of urban planning, the comparison between the shares of road space devolved to the different modes of travel and its rapprochement with the proportions of the population involved in each corresponding mobility aim to symbolically question the space accorded to each mobile individual in public spaces. Sometimes reduced to an opposing of the place given to cars against that given to pedestrians, these distributions have no direct operational meaning: the roadway is also used by modes of public transportation, the sidewalk is not only used by mobile individuals (e.g., for garbage storage, bulky items, street lamps), cars can have several occupants, etc. It is their heuristic scope, more than practical use, that has contributed to reflections on the hierarchies at work on roads.

The use of roads also generates nuisances that align with the dynamics of residential segregation, particularly by making homes in exposed areas relatively less expensive. The health effects of air pollution generated by road traffic, long documented in medical literature, have often been identified through broad statistical findings rather than individual experiences. However, the case of Ella Adoo-Kissi-Debrah, who died at the age of nine in 2013, following severe asthma attacks, brought these impacts into spotlight. For the first time in the United Kingdom, the coroner concluded, after a lengthy judicial procedure, that air pollution had played a role in her death due to her repeated exposure to high levels of pollution, which had tragically been undiagnosed in her lifetime. Ella lived in Lewisham, a working-class neighborhood in South-East London, just 25 m from the South Circular Road, where – as in many cities around the world – levels of fine particulate matter and nitrogen dioxide regularly exceed the limits set by state as well as World Health Organization recommendations (Mandard 2021). As health impacts of roads become more visible, affecting disproportionately working-class neighborhoods, they may be given greater consideration in public policies.

A second register of separation and interdependence operates at the scale of the roadway itself, in terms of space sharing, previously mentioned in the context of gender mainstreaming. This spatial distribution concerns the different mobile places that are the motor vehicles themselves, with the creation of lanes reserved for those carrying multiple passengers (carpooling, buses, etc.), as increasingly for electric vehicles, as seen in Canada or California. Given the higher purchase cost of electric vehicles, which are typically used by middle- and upper-class motorists, this environmental prioritization is therefore also socially exclusionary. Vehicles used by lower income individuals tend to be lighter and thus less polluting, but they are often older and, and for these reasons, more heavily penalized, both in environmental policies and in the event of accidents. Indeed, their lighter weight offers less protection compared to the heavier vehicles often driven by wealthier individuals. This asymmetry helps explain that 22% of fatal accident victims in mainland France and Corsica were blue-collar workers in 2007, despite them comprising only 14% of the population aged 15 years and over (Grossetête 2012; Demoli 2015).

Beyond the separation and hierarchy between the spaces of the roadway, the accidents that occur on them reveal stark inequalities. Class and racial discrimination issues are evident in what Schmitt (2020) calls the "silent epidemic of pedestrian deaths" in the United States. In 2021, car accidents claimed nearly 7,500 lives, the highest number in 40 years and more than double the mortality rate in 2010 (GHSA 2022). Amid indifference to this "traffic violence", Schmitt highlights three contributing factors: increasing road traffic, the growing size of vehicles – especially SUVs – and the residential mobilities of working-class, non-white populations, from urban centers to more car-centric suburban neighborhoods, unfavorable to pedestrians. Afro-American and Hispanic men are twice as likely to be killed while walking than white men. The most dangerous places are high-speed suburban roads with many lanes, passing through lower income neighborhoods, lacking pedestrian crossings and yet located near pedestrian-generating flows places such as interchange hubs, employment centers or shopping areas. Roosevelt Boulevard in North Philadelphia is one such example, particularly in Feltonville, where 73% of residents are Hispanics or Afro-Americans. Spanning 12 lanes and 90 m wide, Roosevelt Boulevard requires pedestrians to cross the equivalent of a soccer field. In 2018, the place was responsible for 21 deaths, nearly one-fifth of Philadelphia's pedestrian casualties. Roadways are not only intermittently marked by pedestrian crossings, but also shaped by segregation dynamics driven by the dominance of automobiles, with potentially grave consequences.

Would actions undertaken in the name of sustainability and environmental transition, which aim to reduce automobile dominance, also address the inequalities driven by road infrastructure at various scales?

9.2.3. *An environmental transition without social justice?*

Public policies aimed at reorienting the development and use of transportation places toward ecological transition risk being socially unjust in three main ways. Each reflect a disjunction that pits the places and actors favored by these actions against those who are marginalized.

A first issue lies in the attempts to marginalize artisanal transportation. Minibuses, auto or cycle rickshaws (otherwise known as cyclos) and other motorcycles play a considerable role in the mobility of large segments of societies in the Global South (as discussed in Chapter 8). However, these modes of transportation are often negatively perceived by decision-makers, who view them as short-lived, place them at the lowest end of the transportation hierarchy and tend to restrict their use, which especially affects vulnerable populations. For instance, the government of Punjab, in Pakistan, proposed to curb the production and use of three-wheeled motorcycles, risking depriving women in rural areas of their main mode of public transportation (Starkey et al. 2021).

Such restrictions can be carried out in the name of transportation sustainability, as Turner (2020) highlighted in Hanoi. Authorities there plan to gradually phase out motorcycles from the city center by 2030, despite them being used by the majority of the city's population. Aspiring to transform the city into a global financial center modeled on Singapore or Hong Kong, they wish to replace them with "sustainable" transportation based on a bus rapid transportation system, launched in 2017, an elevated subway system, which remains largely unfinished. This strategy, implemented under an authoritarian, centralized, communist regime, threatens the daily lives of thousands of motorcycle drivers, known as *xe ôm*, who wait for their customers near train stations and bus stops, hospitals, multiple street crossings and beyond (Figure 9.2). In 2017, a "Clean up the Sidewalks Campaign" was launched, targeting both street vendors and motorcycle parking. It exposed drivers to police fines, especially in the vicinity of markets and tourist attractions, where *xe ôm* frequently wait for their customers. This repression of *xe ôm* affects the "social infrastructure" of working-class Hanoi, that is, the assemblage of social and material relations that enable its existence (McFarlane and Silver 2017). These forms of marginalization further accentuate the vulnerability of artisanal transportation workers, who already endure harsh working conditions (Doherty 2017), including risks of exploitation, traffic accidents (sometimes fatal) and exposure to high levels of pollution. These actions clearly lack social sustainability.

Figure 9.2. *Motorcycle driver in Hanoi (xe ôm). Recent measures limit authorized waiting spaces to marginalize the activity (source: Phan Minh Tuấn (2015))[6]. For a color version of this figure, see www.iste.co.uk/fretigny/transportation.zip*

The spatially selective nature of actions deployed in certain places to the detriment of others constitutes a second form of injustice. A striking illustration is

6 Creative Commons license, available at: https://commons.wikimedia.org/wiki/File:Xe_%C3%B4m,_Li%E1%BB%85u_Giai,_H%C3%A0_N%E1%BB%99i_001.JPG.

the dualization of transportation between urban centers, where alternatives to individual car use are more prominent, and urban peripheries, where automobile dominance prevails, as shown by Pineda and Vogel (2014) regarding Copenhagen since the 1990s. Pedestrianization projects on the streets of urban centers, often motivated by objectives of economic attractiveness (commercial or even tourist-based), cause a reallocation of automobile traffic, reinforced beyond the pedestrianized neighborhoods. The risk is that the associated nuisances, whether from transit traffic or the resulting increased visits to urban centers, become concentrated in the surrounding areas. Combined with the rise of short-term paid car parks near city centers and new surrounding road infrastructures, such as ring roads, pedestrianization projects are not necessarily socially and even environmentally sustainable (Brenac et al. 2013; Reigner and Brenac 2021). More multiscalar politics of mobilities are needed to address these interdependencies.

A third type of injustice lies in the socially selective nature of the actions implemented in the name of transition, in particular in favor of cycling or walking. Regarding cycling, through three case studies in Portland, Milwaukee and Minneapolis, Hoffmann (2016) highlights the over-representation of white, middle-class and upper-class men in cycling associations, events and pro-cycling planning processes. She points to a triple social divide based on class, gender and racial discrimination within cycling cultures in the United States. As a result, cycle paths can then become, in the eyes of African-American communities, "white lanes", contributing to exclusionary dynamics. In Minneapolis, Mayor Raymond T. Ryback pursued a sustainable mobility strategy to boost economic attractiveness and attract members of Richard Florida's (2002) so-called "creative class". He extended the Midtown Greenway, opened in 2000, which runs parallel to the main streets of the city on former railroads, spanning 9 km. In 2010, he introduced a bike-sharing system (Nice Ride), the first of its kind in the United States. However, the planned extension of the greenway into the northern part of the city in 2011 was not welcomed. These working-class neighborhoods, home to many African-American residents, had a low uptake of cycling and lacked bike-sharing stations when Nice Ride was launched. The risk is that the greenway functions as a vector of gentrification, driving up property prices and gradually displacing the original inhabitants. These tensions call for a broad rethinking of the planning processes of places where active mobilities are encouraged.

When it comes to walking, the walking bus, a system of school pickup, while still not commonplace, has been spreading since the 1990s across the Global North and has also proved socially ambiguous. Pigalle (2018), who studied it in the Swiss city of Lausanne and in the French city of (where it is named *pédibus*), shows the normative nature of the discourses underpinning its implementation, stigmatizing parents who are unable to comply with the constraints of the scheme. In particular, it requires parents to be available at the scheduled time to escort their children to the walking bus stop. Upper-class mothers, especially those who work part-time, tend to be more involved in the scheme and often serve as volunteers accompanying the children. Parents from other social classes are more constrained by fixed working hours, making it more difficult for them to participate in the scheme. It is therefore not surprising that the schemes studied are more common in wealthy neighborhoods. The existence of the walking buses can have the adverse effect of ignoring the need to explore alternative actions better suited to the diversity of situations. More broadly, while the deployment of environmental actions is crucial, achieving a socially just transition in transportation and its places is, by all accounts, a considerable challenge.

Practices of transportation places, whether walking bus stops, airplanes or roads, refract various social stratification dynamics. This multidimensionality complicates the dichotomy emphasized in the framework of splintering urbanism where some transportation places and premium networks are dedicated to the few while others suffer the consequences and rely on different infrastructures (Graham and Marvin 2001). A key element in this regard is that differentiations occur not only between transportation networks, but also on a finer scale, within the micro-geographies and micro-politics of the transportation place itself.

9.3. The transportation place as a material and symbolic distancing device

The transportation place can be conceived as a spatial device (*dispositif*), that is to say, "a resolutely heterogeneous set, comprising discourses, institutions, architectural arrangements, regulatory decisions, laws, administrative measures, scientific statements [...], in short: what is said, as much as what is unsaid" (Foucault 1994, p. 299, our translation), through which the material and symbolic place of different actors, whether travelers or others, is played out.

9.3.1. *The kaleidoscopic "place struggle"*

The social and spatial position of passengers in transportation places is the subject of assignments and negotiations involved, literally, a "place struggle" at work in societies, as discussed by Lussault (2009). The kaleidoscopic nature of the situations can be understood by examining the multiple forms of domination at play.

Racial discrimination arguably provides the most striking illustration of micro-scale segregations of the transportation place with Rosa Louise McCauley

Parks' protest on December 1, 1955. In Montgomery, Alabama, the front rows were reserved for white passengers (who had far more seats than their proportion among passengers), while the back rows were allocated to black passengers. The middle seats could be used by black travelers, but only until a white passenger arrived. When Rosa Parks refused to comply with the driver's request to give up her seat to a white passenger in the intermediate rows, several months after similar refusals by others, notably Claudette Colvin – arrested at the age of 15! – her arrest triggered a boycott of the bus company and a broader national mobilization for civil rights. The US Supreme Court abolished bus segregation in November 1956 – after having upheld it in 1896 – but only in 1964 for buses crossing a state border. The micro-geography of Montgomery's buses clearly played the role of reproducing a racist social order and became a matrix of its contestation, rooted in this transportation place while resonating across the much wider scale of US society.

The division into classes – both social and travel-related – can also be very marked in transportation places, echoing the market segmentation of transportation operators. The introduction of travel class distinctions on trains, made compulsory in France by the 1842 Railway Act, was not straightforward. It faced strong opposition in the press, with the satirical daily *Le Charivari* pointing out its inequity (Sauget 2009). Air transportation places demonstrate particularly well this search for differentiation. As Bowen notes (2010, p. 11): "In general, transportation geographers have neglected class; yet the social stratification of transportation systems is not only enduring, in the case of commercial aviation at least, it seems to [be] deepening". The introduction of different airline classes over time reflects the increasing compartmentalization of long-haul flight cabins. The late 1940s saw the creation of the "tourist class" in the United States (now known as economy), which contrasted with what then became "first class". "Business class" emerged in the late 1970s, followed by "premium economy" in the early 1990s.

From the perspective of mobility studies and political studies, fast-track air transportation circuits enjoyed by passengers in the "front classes" or those who have high loyalty points (treated as "frequent fliers") have been widely mentioned, particularly in relation to priority access to passport control. In political and social geography, scholars have advanced the idea of a "business class citizenship" (Sparke 2006) and the concept of a "kinetic elite", an elite of speed, as part of a growing social bipolarization of globalized societies (Sloterdijk 1989; Cresswell 2006).

The notion of kinetic elite, and even more so of hypermobility, raises several points of discussion. First, it assumes a direct link between the intensity of air travel and one's own social position – a connection that is not, however, self-evident (Frétigny 2015). Next, it tends to conflate the market-driven categories allowing the prioritization of some passengers with the social categories stratifying our societies.

The travel industry is highly invested in portraying these profit-driven categories as social distinctions, aiming to increase their perceived value and social acceptance, in the hope that they eventually become social categories. These categories are, in essence, performative – designed to be socially performed by individuals (whether passengers or staff) to reinforce their legitimacy and make them more convincing. Lastly, the prioritization of passage through transportation places relates to diverse logics and hierarchies (front class passengers, frequent passengers, paying users of VIP services at airports, holders of diplomatic passports, etc.), leading to a reconsideration of the homogeneity of populations encompassed by the notions of kinetic elite and hypermobility.

These segregation logics are nonetheless pervasive in the organization of passenger routes, which have been designed to neutralize interactions with other users of the airport public space. The designers, for example, took inspiration from the model of the virtual private network[7]: “a tunnel, a separate space leading right up to the airplane, free of any disturbances, a terminal within the terminal”, as described by an Air France airline executive. These very elaborate routes can include reserved car parks, private elevators, direct car access to the airplane and so on. These devices also provoke certain forms of resistance or derision. In Dubai, police officers do not consider the type of line (priority or not) related to the passport checks they carry out, much to the dismay of the airport operator. At Paris-Charles-de-Gaulle airport, agents jokingly refer to priority passengers as “feather dusters”, a metonym for the plastic plants placed at the edge of the priority lines, meant to reinforce the distance, both material and symbolic, from other passengers (Figure 9.3). But mistakenly starting to join the priority line and being turned away can trigger strong emotions (embarrassment, shame, etc.), revealing the full force of these devices (Frétigny 2013).

Opposed to these accelerated circuits are those that slow down or immobilize travelers, following logics of nationality and migration status. While these migratory logics permeate multiple places, they have a considerable influence on transportation places functioning as border nodes. Slowing down is noticeable in the passport control lines for non-priority nationalities, such as non-Gulf Cooperation Council nationals at airports within these states. More dramatically, they come to a halt in migration detention centers such as the “holding areas for persons whose cases are pending” (ZAPI) in France. These detention centers can be temporary or permanent, as in the case, for the latter, of ZAPI 3, located at the foot of the Paris Charles-de-Gaulle airport runways, between terminals 1 and 2. They hold passengers without

7 A system of exchange between remote computers where communications occur via public networks but are carefully separated from other traffic.

authorization to access French territory, either because they are seeking asylum and awaiting a preliminary examination of whether their application is "manifestly unfounded" or because, during transfer at the airport, they were not authorized to board their departing plane. The conditions of detention in such centers are regularly criticized by aid organizations. In April 2021, the Red Cross and the National Border Assistance Association for Foreigners (ANAFÉ) decided to withdraw temporarily from ZAPI 3 to protest against the sharp increase in the number of detainees. The number rose that year from 70 in late March to 133 on April 27, during the third wave of Covid-19, accompanied by national lockdown measures. These NGOs decried a lack of testing, poorly ventilated premises and the material impossibility of applying health protocols in the facility (Birchem 2021). While these contemporary places are far less known than Ellis Island[8], the main gateway – for admitted migrants – to the United States via the Port of New York from 1892 to 1954, they play today in various respects a comparable role.

Figure 9.3. *Fast-track access at Paris-Charles-de-Gaulle airport (source: Frétigny (2011)). For a color version of this figure, see www.iste.co.uk/fretigny/transportation.zip*

These migratory procedures closely intersect with gender relations, potentially resulting in the separation and restriction of migrant women's agency. Indonesian

8 Other transportation places such as international railway stations have played an early key role in border controls and deportation practices, from the mid-19th century, as shown by Torsten Feys (2019) in relation to Belgium.

migrant domestic workers returning to Jakarta's Soekarno-Hatta International airport are identified and set apart by agents upon arriving at the airport, as part of a government procedure intended to limit extortion attempts targeting them as bearers of remittances (Kloppenburg and Peters 2012). They are taken to a special terminal, guarded and off-limits to third parties, then transferred in minibuses of government-designated companies. Drivers are required to drop them off at the front door at the address featuring on their passport, against the signature of a family member. This system prevents them from choosing an alternate address, unless their husband or parents reside there. Presuming the vulnerability of migrant women, it exposes them to informal but imperative payments demanded by drivers aboard minibuses. These payments have the adverse effect of helping to fuel the rights of passage levied by gangsters on small access roads to villages, illustrating the great complexity of power dynamics within this device.

Figure 9.4. *Barrier limiting access to part of the subway platform for women and children under 12, in Mexico City (source: Frétigny (2018)). For a color version of this figure, see www.iste.co.uk/fretigny/transportation.zip*

More broadly, the issue of segregation within transportation places, framed as protecting a social group, arises. Such separations frequently appear in the form of women-only compartments or cars on subways or Bus Rapid Transit networks in Dubai, Mexico City (Figure 9.4), Brazil, Cairo, Iran, Kuala Lumpur, Manila, Taipei, Seoul and Japan (Tillous 2016). In Cairo, the 1989 decision to create women-only cars, also accessible to boys under 12 and people with disabilities, was not aimed at preventing harassment. The objective of decision-makers fell within the field of traffic management: reducing road congestion by providing women with a level of "comfort" on the subway comparable to that of a car, encouraging them to shift transportation mode. The decision was also rooted in a religious, morally conservative perspective concerning the capital's subway, which opened in 1987.

Designers saw the subway as a space meant to shape people's behavior by increasing respect for public morality, with sexual harassment being condemned from this perspective. Feminist activists in Cairo focused on other battles, at least until the early 1990s. This time marked a turning point in favor of fighting against sexual harassment. Its introduction into the public agenda took place within the broader framework of human rights and women's rights on a global scale, pushed by international organizations. The highly publicized sexual assault of a young woman boarding a bus in broad daylight at Attaba Square in 1992, in central Cairo, may well have consolidated the policy of separate cars on the subway. Faced with the multiple questions and motivations that this type of separation brings into play, Tillous (2017) points to the convergence of positions of two contemporary feminist associations, Bassma and ShoftTa7rosh: "The women-only car is not a solution, and if it did not exist, it would not need to be introduced. But since it does exist, it is considered a lever for action against harassers and a way to avoid the risk of harassment, in the absence of better options". This captures both the specificity and the multifaceted nature of gender-based separations within transportation places.

The separations at work within transportation places thus gain from being grasped in all their complexity. In particular, the practices of transportation places are differentiated by population categories over time. This temporal segregation can be observed for instance through hourly variations during a working day in the use of a public transit line by commuters, depending on their working hours, which are closely tied with their professional status. It arises more strikingly in times of disruption, such as during the early days of the Covid-19 pandemic, in 2020. When comparing 2020 data to those of 2019, the crisis provoked a sharp decline in the use of subway station turnstiles in wealthier, western Paris neighborhoods (Finance 2021). In contrast, the drop in the number of passengers at stations in eastern Paris, which is more working-class, was more moderate: a 35% decrease at Château Rouge for instance, compared to an average decrease in Paris of 49%. The capacity, or not, to work remotely and, more broadly, to avoid using the subway played a major role in these disparities.

Whether we consider relations of race, social or travel class, nationality, migratory status or gender, passengers occupy a major place in the spatial and temporal dynamics of distancing at stake in transportation places. However, many other stakeholders are involved as well.

9.3.2. *Beyond passengers*

Beyond and often in the immediate vicinity of passengers, transportation workers and other users are fully engaged in distancing practices. Some of those serving the mobility of others, often in very specific locations, are exposed to arduous working conditions or forms of exploitation. This is what led Cresswell (2006) to refer to them as members of a "kinetic underclass". These experiences are rarely conspicuous in places designed for passengers. This poster from Transport for London (Figure 9.5), the London public transit operator, is an exception in this respect, by portraying the emotions and suffering that may occur behind the uniform.

The rise of ride-hailing services and the gig economy have intensified the challenging working conditions playing out on the other side of the vehicle. But these conditions were already previously identified. Sharma (2008) documents, for example, how Chicago taxi drivers experience stress transmitted by passengers worried about missing their flight on the way to the airport. She also highlights the joint pain drivers endure from being confined to their seat, their irregular schedules ("cab lag" as opposed to jet lag) and the difficulty of accessing bathrooms, which forces them to give up drinking coffee. These experiences behind the wheel of North American taxis are shaped by gender and race relations, as most drivers are male immigrants, with 70% identifying as South Asian in New York and Toronto.

The gendered bodies of workers are subject to intense social control in transportation places, with flight attendants being one of the most emblematic examples. Singapore Airlines female flight attendants fully illustrate this gendered bodily discipline due to their role in the airline's marketing strategy. Created by the Ministry of Finance in 1972, which sees it as the bridgehead of Singapore's insertion into international relations and world trade, the city-state's flagship company banked on the "Singapore girl" to represent it from its beginnings. This iconic figure of the female flight attendant is responsible for embodying in-flight service quality and a nurturing atmosphere[9] that enhances passenger comfort, setting the company apart from competitors, both Western airlines and from nearby countries such as Malaysia or Vietnam (Lin 2015). By drawing on Western stereotypes of exoticism, gender and "Asian hospitality" (a term that often appears in the company's discourses aimed at female flight attendants), the company promotes a form of self-Orientalism,

9 On this concept, see Chapter 7.

submission and eroticization of the female body. The company has thus earned symbolic recognition, reflected, for example, in 1993 by the introduction to Madame Tussauds' London Museum of the "Singapore Girl", the first commercial figure to ever appear there.

Figure 9.5. *Serving the mobility of others as a grueling experience: poster in the London Underground (source: Frétigny (2022)). For a color version of this figure, see www.iste.co.uk/fretigny/transportation.zip*

The airline's gray literature, available up until the early 1990s and studied by Lin, shows how much female flight attendants, expect to embody this ideal through their own performance, have been and still are subjected to considerable physical and emotional labor, that North American and European crews, also subjected to sexist treatments, were able to contest to a greater degree from the 1970s. This labor focused on their makeup and silhouette and indirectly their lifestyle, gestures and even speech, as Singaporean accents were smoothed out through diction lessons. The airline also fostered extensive affective engineering to encourage self-discipline in courtesy, for example, through annual sketch competitions that highlighted its importance. These trainings were aimed at differentiating from competitors by refining interactions with economy-class passengers, mothers with children (fathers were not mentioned), passengers on their way to a funeral, etc. Practices and interactions in the vehicular place that is the cabin thus proved subject to very strict standards of behavior.

The social hierarchy of aviation workers is in many ways tied to their spatial proximity to passengers and whether or not they are in contact with them. Duchêne (2011), in an ethnographic study of a company managing baggage and passengers at Zurich airport, points out that the most visible agents are the ground-handling staff, in charge of check-in, boarding, etc. Most of them are Swiss nationals. Semi-visible workers, responsible in particular for guiding passengers with disabilities through the airport, are often either Swiss, from working-class backgrounds, or migrants. It is also the latter who occupy the bulk of working positions behind the scenes, such as baggage handlers, in the technical areas of the baggage sorting system, and outside, on aircraft parking areas. Their language skills are not recognized, nor are their qualifications in this field. However, the company catalogues their proficiency in rare languages, so that they can serve as interpreters during interactions with passengers who do not speak the languages learned by ground-handling staff. When their assistance is necessary, they are called upon to intervene in the public space of the airport, without this interpreting activity earning them any financial recognition. This underscores how transportation places, as production sites, are heavily marked by inequalities.

The management policies of transportation places tend to marginalize other users, such as homeless or exiled people, who engage with these resource places more at their spatial and temporal margins. The trees on the forecourt of the Gare de l'Est train station in Paris come to mind, where Afghan exiles, often having traveled from Germany, and thus arriving via this station, had placed their belongings in plastic bags resting on the branches during the day, to retrieve them in the evening. The artist Francisca Benitez explored these practices by creating a series of photographs ("*Garde de l'Est*") highlighting the trees' unexpected function in July 2005. The precarious appropriations of the surroundings of the station by Afghan exiles, varying at different time scales, also extended to the Villemin garden, adjacent to the station on one side and the Saint-Martin canal on the other. The usage of this garden as a place to camp in the migratory journey of exiles became particularly visible after the closure of the Sangatte accommodation center, near the Channel Tunnel, in 2002, despite the occupation of such places being subject to regular dismantling interventions. These practices form part of a certain continuity in the use of the Villemin garden as a place of transit for soldiers mobilized during both World Wars as well as for the many displaced persons in the aftermath of these conflicts (Paté 2017). Associations such as the collective supporting exiles in the 10th arrondissement of Paris, created following the use of another green space near the station, the Alban-Satragne Square, have sought to influence public authorities to ensure proper care for these exiles. In contrast, the treatment of Ukrainian exiles in March 2022, following the Russian invasion of February, with the temporary installation of a Red Cross reception center in a station café (Lavignon 2022), as in other main European transportation places, stood out as an exception. Through the spatial devices of transportation places, it is the relationship of each and every one to the city and society that is played out.

9.4. Conclusion

Transportation and its places play a major role in the processes of mobile segregation, through the contrasting frequentation of transportation places and the position devolved to these different places and their actors, including nearby residents, in social hierarchies. These dynamics are also at play through micro-political differentiations, within the spatial device formed by the transportation place itself.

In the situations examined, the distancing of people undoubtedly places minorities in an unfavorable position. However, this does not prevent those subjects to power relations from seeking themselves greater distance in the travel experience in specific circumstances, especially in response to harassment. Previously discussed in relation to women-only cars or compartments, the lack of distance can prove particularly concerning and lead to the avoidance of places of public transportation. Resorting to costly alternatives, such as taxis, or taking long detours, is a recurring mobile practice, especially among women or LGBTQ+ people (Weintrob et al. 2021).

While in this chapter the focus has been placed on separation, transportation places prove eminently ambivalent, as they can also play an inclusive role as public, private or hybrid sites. Through multiple ethnographic observations in Great Britain, Laurier et al. (2008) have demonstrated that the car is a considerable place for the deployment of social life, especially in family, friendship and professional life that unfold within the vehicle. In an entirely different context, in Mumbai and other Indian cities, particularly near railway stations, the street edges, ranging from part of the sidewalk to the limits of the carriage-way, are invested as domestic space, with the help of bamboo poles and fabrics. Banerjee (2022) invites us to revisit the categories used to describe these situations. He stresses that these "footpath-dwellers" fully consider them as their own home (*ghar*), without, in this conception, the private appropriation of this street edge being contradictory to the larger public space status of the street, with all the associated social interactions. Cycling schools, where adults, particularly immigrant women, learn to ride in working-class neighborhoods, are places that foster the emancipation and sociability of these women (Frétigny et al. 2024). They clearly illustrate the challenge of "commoning mobilities" (Nikolaeva et al. 2019) and communing transportation places in the face of crucial social and environmental issues.

The integrative contribution of transportation places is also of a symbolic and memorial nature. Public art in these places can celebrate the contribution of minorities, such as the bronze sculpture installed in 2022 at Waterloo Station in London, named the National Windrush Monument. It commemorates the highly publicized arrival of Afro-Caribbeans aboard the HMT *Empire Windrush* in 1948. These ship passengers have become emblematic of the nearly 500,000 Afro-Caribbean migrants who came to the United Kingdom after the war to provide their labor force, referred to nationwide as the Windrush Generation. Near the sculpture, a poem by Laura Serrant is addressed to the passengers: "You Called … and We Came". This recognition proves all the more significant given that members of the Windrush generation were wrongly detained, threatened with deportation and in some cases deported in the 2010s, sparking a scandal, known as the Windrush Scandal, in 2018 when these situations were brought to public attention. Such cultural actions aim to ensure that transportation and its place repair, as far as possible.

Moreover, the separations at stake in transportation places are also subject to appropriations and objections to have them evolve in a positive direction, if only temporarily. We may think of the public transportation desegregation movement launched in the wake of Rosa Parks' action but also the disability pride movements. Starting in 1990 in the United States and drawing from the same foundational ideas as gay pride movements, they cast light on and denounce, in the street space, the disabling situations engendered by exclusionary transportation places. In similar forms, cyclists counter their marginalization by showing the "critical mass" they can achieve – Critical Mass being the name of this type of cyclist mobilization – in multiple cities around the world, where they gather en masse on weekends to block automobile traffic and enjoy riding, turning the route into a form of protest.

Objections also takes the form of memorial work, such as that undertaken by African-American communities whose neighborhoods have been disrupted by the construction of urban highways. Unlike the victorious and widely publicized battle in the 1960s led by urban planner Jane Jacobs to protect Greenwich Village from an expressway in New York, many non-white communities, despite their mobilization, could not prevent its construction. The Rondo neighborhood, the most symbolic of the Afro-American community in the Minneapolis-Saint Paul metropolitan area, has commemorated since 1983 the lost urban spaces and the disruption caused by the creation of Interstate Highway 94 in the mid-1960s, through a festival and a parade, known as Rondo Days. The objective of such initiatives, as analyzed by Avila (2014), is to reclaim highways and their surroundings as performance sites blending political practices, identity expression and the quest for well-being. Overcoming mobile segregation, in which transportation sites play a key part, is a vast undertaking that begins by acknowledging its scope, considering the multiplicity of sites, scales, actors and processes involved.

9.5. References

ART (2022). Rapport annuel sur le marché du transport par autocar et sur les gares routières en France en 2021 [Online]. Available at: https://www.autorite-transports.fr/wp-content/uploads/2022/07/rapport-sur-le-marche-du-transport-par-au-tocar-et-les-gares-routieres-2021.pdf [Accessed 27 October 2022].

Avila, E. (2014). *The Folklore of the Freeway: Race and Revolt in the Modernist City*. University of Minnesota Press, Minneapolis.

Birchem, N. (2021). Dans la zone d'attente de Roissy, le Covid inquiète. *La Croix* [Online]. Available at: https://www.la-croix.com/France/zone-dattente-Roissy-Covid-inquiete-2021-04-28-1201153079 [Accessed 5 November 2022].

Bowen, J. (2010). A people set apart: The spatial development of airline business class services. In *International Business Travel in the Global Economy*, Beaverstock, J., Derudder, B., Faulconbridge, J., Witlox, F. (eds). Ashgate, Farnham.

Brenac, T., Reigner, H., Hernandez, F. (2013). Centres-villes aménagés pour les piétons : développement durable ou marketing urbain et tri social ? *RTS – Recherche Transports*, 29, 271–282.

Budd, L. and Hubbard, P. (2010). The 'Bizjet set': Business aviation and the social geographies of private flight. In *International Business Travel in the Global Economy*, Beaverstock, J.V., Derudder, B., Witlox, F. (eds). Ashgate, Farnham.

Cresswell, T. (ed.) (2006). The production of mobilities at Schiphol Airport, Amsterdam. In *On the Move: Mobility in the Modern Western World*. Routledge, London.

Demoli, Y. (2015). Carbone et tôle froissée. L'espace social des modèles de voitures. *Revue française de sociologie*, 56(2), 223–260.

Demoli, Y. and Subtil, J. (2019). Boarding classes. Mesurer la démocratisation du transport aérien en France (1974–2008). *Sociologie*, 10(2), 131–151.

Doherty, J. (2017). Life (and limb) in the fast-lane: Disposable people as infrastructure in Kampala's boda boda industry. *Critical African Studies*, 9(2), 192–209. doi: 10.1080/21681392.2017.1317457.

Duchêne, A. (2011). Néolibéralisme, inégalités sociales et plurilinguisme : l'exploitation des ressources langagières et des locuteurs. *Langage et société*, 136(2), 81.

Emery, J., Motte-Baumvol, B., Hivert, L. (2021). Motifs de démotorisation des actifs et retraités dans les aires urbaines françaises : premiers résultats statistiques de l'en-quête MoDe. *RTS – Recherche Transports Sécurité*. doi: 10.25578/RTS_ISSN1951-6614_2021-01.

Fainstein, S.S. (2014). The just city. *International Journal of Urban Sciences*, 18(1), 1–18.

Feys, T. (2019). Riding the rails of removal: The impact of railways on border controls and expulsion practices. *The Journal of Transport History*, 40(2), 189–210.

Finance, O. (2021). Le métro parisien en période pandémique : le privilège de l'immobilité. *Cybergeo Conversation* [Online]. Available at: https://cybergeo.hypotheses.org/964 [Accessed 24 September 2022].

Florida, R.L. (2002). *The Rise of the Creative Class: How It's Transforming Work, Leisure, Community and Everyday Life*. Basic Books, New York.

Fol, S. and Gallez, C. (2017). Évaluer les inégalités sociales d'accès aux ressources : intérêt d'une approche fondée sur l'accessibilité. *Revue Internationale d'Urbanisme*, 4 [Online]. Available at: http://www.riurba.review/Revue/evaluer-les-inegalites-sociales-dacces-aux-ressources-interet-dune-approche-fondee-sur-laccessibilite/.

Foucault, M. (1994). *Dits et écrits, III. 1954–1988*. Gallimard, Paris.

Fraser, N. (1996). Social justice in the age of identity politics: Redistribution, recognition, and participation. In *The Tanner Lectures on Human Values*. Stanford University, Stanford.

Frétigny, J.-B. (2013). Les mobilités à l'épreuve des aéroports : des espaces publics aux territorialités en réseau. Les cas de Paris Roissy-Charles-De-Gaulle, Amsterdam Schiphol, Francfort-sur-le-Main et Dubai International. PhD Thesis, Université Paris I Panthéon-Sorbonne, Paris.

Frétigny, J.-B. (2015). L'hypermobilité : une figure introuvable ? *Sociologies* [Online]. Available at: https://sociologies.revues.org/5166 [Accessed 27 November 2015].

Frétigny, J.-B., Bouloc, C., Bocquillon, P., Cazeaux, L. (2024). *Décarboner les mobilités : quelles politiques en France ?* Éditions de la Sorbonne, Paris.

Gallez, C. (2023). Mobility justice as a political object. In *Mobility and Geographical Scales*, Drevon, G. and Kauffman, V. (eds). ISTE Ltd, London, and John Wiley and Sons, New York.

GHSA (2022). Pedestrian traffic fatalities by state: 2021 preliminary data. Report, Washington, D.C.

Gössling, S. and Humpe, A. (2020). The global scale, distribution and growth of aviation: Implications for climate change. *Global Environmental Change*, 65, 102194. doi: 10.1016/j.gloenvcha.2020.102194.

Graham, S. and Marvin, S. (2001). *Splintering Urbanism: Networked Infrastructures, Technological Mobilities and the Urban Condition*. Routledge, London.

Grossetête, M. (2012). *Accidents de la route et inégalités sociales : les morts, les médias et l'État*. Éditions du Croquant, Bellecombe-en-Bauges.

Guillot-Wright, S. (2021). "The company will fire you because you are too expensive": A photo-ethnography of health care rights among Filipino migrant seafarers. *Humanities and Social Sciences Communications*, 8(1), 1–10. doi: 10.1057/s41599-021-00947-y.

Harvey, D. (2001). *Spaces of Capital. Towards a Critical Geography*. Routledge, New York

Hopkinson, L. and Cairns, S. (2021). Elite status: Global inequalities in flying. Report, Possible, London [Online]. Available at: https://policycommons.net/artifacts/1439908/elitestatusglobalinequalitiesinflying/2067509/.

Imrie, R. (2001). Barriered and bounded places and the spatialities of disability. *Urban Studies*, 38(2), 231–237.

ITF (2022). Current registries listed as FOCs [Online]. Available at: https://www.itfseafarers.org/en/focs/current-registries-listed-as-focs [Accessed 26 October 2022].

Jackowska, O. and Novas Ferradás, M. (2022). Who owns public spaces? The trailblazer exhibition on women's everyday life in the City of Vienna (1991). *Planning Perspectives*. doi: 10.1080/02665433.2022.2074526.

Kloppenburg, S. and Peters, P. (2012). Confined mobilities: Following Indonesian migrant workers on their way home. *Tijdschrift voor economische en sociale geografie*, 103(5), 530–541.

Laurier, E., Lorimer, H., Brown, B., Jones, O., Juhlin, O., Noble, A., Perry, M., Pica, D., Sormani, P., and Strebel, I. (2008). Driving and "passengering": Notes on the ordinary organization of car travel. *Mobilities*, 3(1), 1–23.

Lavignon, B. (2022). Réfugiés ukrainiens accueillis à la gare de l'Est : "Le système monte en pression et sature". *Libération* [Online]. Available at: https://www.liberation.fr/societe/refugies-ukrainiens-accueillis-a-la-gare-de-lest-le-systeme-monte-en-pression-et-sature-20220317_3C4PLNAMDJAOZGKXAP42TN2DEM/ [Accessed 16 November 2022].

Lin, W. (2012). Wasting time? The differentiation of travel time in urban transport. *Environment and Planning A*, 44(10), 2477–2492.

Lussault, M. (2009). *De la lutte des classes à la lutte des places*. B. Grasset, Paris.

Mandard, S. (2020). Ella Adoo-Kissi-Debrah, 9 ans, "tuée" par la pollution de l'air. *Le Monde* [Online]. Available at: https://www.lemonde.fr/planete/article/2020/12/17/ella-adoo-kissi-debrah-9-ans-tuee-par-la-pollution-de-l-air_6063670_3244.html [Accessed 31 October 2022].

McFarlane, C. and Silver, J. (2017). Navigating the city: Dialectics of everyday urbanism. *Transactions of the Institute of British Geographers*, 42(3), 458–471. doi: 10.1111/tran.12175.

Mobiliscope (2023). La ville à toute heure. Les populations présentes et leurs mobilités au cours de la journée [Online]. Available at: https://mobiliscope.cnrs.fr/fr [Accessed 13 January 2023].

Nikolaeva, A., Adey, P., Cresswell, T., Lee, J.Y., Nóvoa, A., and Temenos, C. (2019). Commoning mobility: Towards a new politics of mobility transitions. *Transactions of the Institute of British Geographers*, 44(2), 346–360. doi: 10.1111/tran.12287.

Passalacqua, A. (2009). L'autobus et Paris. Souplesse, espace public et mobilité de 1900 aux années 1970. PhD Thesis, Université Paris Diderot, Paris.

Paté, N. (2017). De l'enfant dans la rue au citoyen dans la rue. *Vacarme*, 4(81), 77–83.

Pigalle, É. (2018). Sur le chemin des écoliers, les politiques en faveur de la marche à pied : le pédibus en France et en Suisse romande : discours, déploiement et pratiques. PhD Thesis, Aix-Marseille, Marseille.

Pineda, A.F.V. and Vogel, N. (2014). Transitioning to a low carbon society? The case of personal transportation and urban form in Copenhagen: 1947 to the present. *Transfers*, 4(2), 4–22.

Ram, E. (2021). How Nairobi's "road for the rich" resulted in thousands of homes reduced to rubble. *The Guardian* [Online]. Available at: https://www.theguardian.com/global-development/2021/dec/08/how-nairobis-road-for-the-rich-resulted-in-thousands-of-homes-reduced-to-rubble [Accessed 31 October 2022].

Reigner, H. and Brenac, T. (eds) (2021). *Les faux-semblants de la mobilité durable : risques sociaux et environnementaux*. Éditions de la Sorbonne, Paris.

Sauget, S. (2009). Les mobilités vues depuis les gares parisiennes avant 1914. In *De l'histoire des transports à l'histoire de la mobilité ?*, Flonneau, M. and Guigueno, V. (eds). Presses Universitaires de Rennes, Rennes.

Schwarzkopf, S. and Backsell, J.I. (2021). The nomos of the freeport. *Environment and Planning D: Society and Space*, 39(2), 328–346.

Sharma, S. (2008). Taxis as media: A temporal materialist reading of the taxi-cab. *Social Identities*, 14(4), 457–464.

Sheller, M. (2018). *Mobility Justice: The Politics of Movement in an Age of Extremes*. Verso, London.

Sloterdijk, P. (1989). *Eurotaoismus: zur Kritik der politischen Kinetik.* Suhrkamp, Frankfurt am Main.

SMTR (2018). Les habitudes de déplacements des actifs et des inactifs. Report [Online]. Available at: https://smtr-mobilite.re/wp-content/uploads/2018/11/SMTR_LIVRET-4-Actifs-inactifs_16283_BD.pdf [Accessed 27 October 2022].

Sparke, M.B. (2006). A neoliberal nexus: Economy, security and the biopolitics of citizenship on the border. *Political Geography*, 25(2), 151–180.

Spence, E. (2017). Beyond the city: Exploring the maritime geographies of the superrich. In *Cities and the Super-Rich: Real Estate, Elite Practices and Urban Political Economies*, Forrest, R., Koh, S.Y., Wissink, B. (eds). Palgrave Macmillan, London.

Starkey, P., Batool, Z., Younis, E.M.W., Rehman, A.U., Ali, M.S. (2021). Motorcycle three-wheelers in Pakistan: Low-cost rural transport services, crucial for women's mobility. *Transportation Research Interdisciplinary Perspectives*, 12, 100479. doi: 10.1016/j.trip.2021.100479.

Tillous, M. (2016). Women-only cars. Cartographie des dispositifs de séparation des sexes dans les réseaux de métro du monde [Online]. Available at: https://metro.hypotheses.org/26 [Accessed 6 November 2022].

Tillous, M. (2017). Les voitures de métro réservées aux femmes comme instrument d'action publique : une réponse à quel problème ? *Géocarrefour*, 91(1). doi: 10.4000/geocarrefour.10033.

Turner, S. (2020). Informal motorbike taxi drivers and mobility injustice on Hanoi's streets. Negotiating the curve of a new narrative. *Journal of Transport Geography*, 85, 102728. doi: 10.1016/j.jtrangeo.2020.102728.

Verlinghieri, E. and Schwanen, T. (2020). Transport and mobility justice: Evolving discussions. *Journal of Transport Geography*, 87, 102798. doi: 10.1016/j.jtrangeo.2020.102798.

Weintrob, A., Hansell, L., Zebracki, M., Barnard, Y., Lucas, K. (2021). Queer mobilities: Critical LGBTQ perspectives of public transport spaces. *Mobilities*, 16(5), 775–791.

Zukin, S. (1995). *The Cultures of Cities*. Blackwell, Oxford.

10

Heritage Development of Transportation Places

Pierre-Louis BALLOT
CITERES, Université de Tours, France

10.1. Introduction

> The goal is as follows: to awaken and reveal the tourist interest and by extension the commercial interest of the old vacation route. This is with a view to making the RN7 no longer simply a place of transit, but a destination in itself. (Hubert Prévost, Deputy Mayor of Pierrelatte (Drome), our translation)

Making a road a destination: An objective that at first glance may seem paradoxical, the road generally constituting a means to reach a destination, a place of residence or of vacation, for example. Yet this is the objective pursued by the association "Aire N7", created in 2012, aiming to group together the 170 municipalities in France crossed by National Road 7 (RN7) in order to promote its route.

Since the 2000s, transportation places have tended to be understood not only based on their functional dimensions, but also through the prism of their heritage dimensions. A process "applying both to an object and to an ideal reality [...], and which is a recognition and development of buildings, inherited spaces, objects and practices" (Fagnoni 2011, our translation), heritage development, applied to transportation places, concerns a variety of objects. These include other routes, such as Route 66 in the United States, for which sightseeing tours are now organized, as

Geographical Places in Transportation,
coordinated by Jean-Baptiste FRÉTIGNY. © ISTE Ltd 2025.

well as ports and stations. For example, the classification as UNESCO World Heritage sites of the commercial port of Liverpool (England) in 2004 and Chhatrapati-Shivaji station in Mumbai (India) in 2004.

The “pragmatist turn” (Clot-Goudard and Tillous 2008) occurring in the 1980s in social sciences initiated a change of perspective on transportation places, from a humanist perspective, under the impetus in particular of the work of sociologist Isaac Joseph devoted to the subway (Joseph 1987) and railway stations (Joseph 1999). However, transportation places are still essentially considered today through the prism of their functional dimensions. In transportation geography, where this approach is predominant, it is the economic and environmental impacts on the territories crossed by these infrastructures that are questioned, as well as the organization of their service and the territorial reconfigurations that they generate (Bavoux et al. 2005).

Works in the social sciences have proposed to go beyond the functional and utilitarian understanding of transportation places from the 2000s, at the very moment when new approaches to mobility were emerging, contributing to the appearance of a “mobility turn“ (Hannam et al. 2006). Thus, in the field of English-language anthropology and ethnology, works dedicated to the road have shown that this infrastructure can “matter to people” (Harvey and Knox 2015), occupying a place as an organizing element of human life throughout history (Snead et al. 2009) and playing a role in creating social interactions and the day-to-day organization of individuals (Meier et al. 2016; Diener and Batjav 2019).

Like these works, entry through heritage development leads to greater interest in the symbolic aspects of transportation places rather than their functionality. Accordingly, this chapter pursues a twofold objective: to describe the different forms of heritage development of transportation places, on the one hand, and to characterize the different issues raised by this process, on the other hand. A generic term, the “transportation place” refers here to the space where the act of transportation is executed, whether it is a specific point (such as an airport) or a route, either physically materialized (e.g. a road) or not (e.g. a flight path), which leads from a point of departure to a point of arrival (Lombard and Steck 2004). Diverse transportation places have been the subject of such a heritage development that their contemporary uses no longer come under a logic of transportation, such as Gare d’Orsay in Paris, which was made a museum in 1986, the Ellis Island complex, off Manhattan, both a ferry terminal and an immigration center from 1892 to 1954, transformed into a museum in 1990, and Berlin-Tempelhof Airport, with its variety of uses as a park, exhibition and sporting event site, etc. While their study is certainly of interest, this chapter opts for a study of transportation places whose functional logics associated with travel still occupy a place in order to be fully able to convey the tension arising with the symbolic aspects of heritage development.

Section 10.2 is devoted to transportation places that are subject to archetypal institutional processes of heritage development, which are characterized in particular by their classification as UNESCO World Heritage sites. Taking the example of stations and ports, the goal is to identify the different criteria advanced to justify this classification and this heritage interest. Then, the subsequent sections propose to focus on more specific forms of heritage development of transportation places, mobilizing the case of roads (section 10.3) and airports (section 10.4). While also returning here to the criteria that justify this heritage interest, we will see that this choice is explained by the fact that these two types of transportation places call on two distinct forms of heritage development, which invite us to reflect successively on the territorial restructuring that takes place via this process (through studying the road) and to reconsider the traditional forms of heritage development (with regard to the airport).

10.2. A traditional heritage development of transportation places: the example of stations and ports

Traditionally, heritage development applies to objects whose actors consider that they deserve to be transmitted from the past to find a value in the present (Lazzarotti 2003), hence the fact that the heritage resulting from this process generally resembles objects that are non-contemporary or that have lost their functional use in the present day. It is partly under this logic that the process of heritage development of certain transportation places, such as stations and ports, operates. In many respects, they play the role of symbol-places, that is to say places that make it possible to render sensory and intelligible on a fine scale the values and representations of a territory and that are often placed at the center and at the limit of these territories (Debarbieux 1995), which corresponds well to the positioning of transportation places.

10.2.1. *Stations and ports: symbol-places of urban, national and imperial territories, and their diverse values*

Since the second half of the 20th century, stations have been the subject of increased heritage interest. This can first be explained by the fact that stations play the role of symbol places. This is the case, for example, with Chhatrapati Shivaji station in Mumbai, listed as a UNESCO World Heritage site since 2004 and the fruit of a dual heritage: colonial, on the one hand, and postcolonial, national, on the other hand. The very name of the transportation place illustrates this desire to move past the logic of British imperial domination, adopted in 1996 in place of its previous name, Victoria Terminus, in reference to the sovereign. Indeed, Chhatrapati Shivaji is the founder of the Maratha Empire, which gave its name to the station's anchor State, Maharashtra. Among the criteria used to justify the listing, experts mainly mentioned the architectural qualities of the station, presented as a "magnificent example of Victorian neo-Gothic architecture in India" (UNESCO 2004), which contributed to it becoming a symbol of the metropolitan area of Mumbai, considered the "Gothic City and the major international mercantile port of India" (UNESCO 2004). But certain elements, such as the stone dome and turrets, are also reminiscent of the traditional architecture of Indian palaces (Figure 10.1).

Figure 10.1. *Chhatrapati-Shivaji station in Mumbai (source: Alexander Savin (Free Art License)[1], 2016). For a color version of this figure, see www.iste.co.uk/fretigny/transportation.zip*

Stations are also testament to the shifts occurring in societies from the mid-19th century. In Europe and North America, they offer markers of the industrialization of national territories. This is the case in France, for example, where this process was enabled by the networking of the country in the first half of the 20th century. These stations therefore form part of the industrial heritage of many cities (Baron and Roseau 2016), such as Paris (Gare du Nord), London (Euston Station) and Liverpool (Lime Street Station). According to historian Georges Duby (1983), the station has

1 Available at: https://commons.wikimedia.org/wiki/File:Mumbai_03-2016_66_Victoria_Terminus.jpg.

constituted, since the end of the 19th century, one of the "major expressions" of the city and the values of societies, like Gothic cathedrals 700 years earlier.

In recent decades, this heritage interest has also extended to railway structures as a whole and in particular to railway lines. Thus, railway lines are established as heritage objects both for their infrastructure (tunnels, viaducts, etc.) and for the significant place that they have held in the networking of a country, associated with values of belief in progress, and in the celebration of the national landscapes crossed. This is the case, for example, with the Semmering railway line (Austria), listed as a UNESCO World Heritage site in 1998 and described as one of the "greatest feats of civil engineering during the pioneering phase of railway building" (UNESCO 1998). Another of the criteria mentioned is the fact that this railway line makes it possible to pass through 41 km of high mountains, considered a "spectacular landscape".

Like stations, ports are also the subject of heritage interest today, both for their role in the development of trade and mobilities on a global scale, and for their architecture. Thus, in 2012, the commercial port of Liverpool was designated a UNESCO World Heritage Site. As a major center of world trade in the 18th and 19th centuries, the city played a significant role in the rise of the British Empire by becoming the main transit point for migratory movements to America. Liverpool also symbolizes the development of modern port technology, its transportation systems and port management. Sometimes ports are integrated into a broader heritage development process, which may concern a territory in which they belong. In the justifications of the UNESCO classification of the Loire Valley in 2000, its traditional ports were cited among the landscape motifs constituting resources for cultural animation.

Thus, if the heritage development of these transportation places is justified in part by the history and territorial representations they embody, the question of their future frequently arises, particularly for stations.

10.2.2. *The future of heritage stations: inherited objects faced with contemporary issues*

While many stations are established as heritage objects, they nevertheless most often retain functional aspects, welcoming numerous passengers daily. Consequently, these stations are regularly the subject of redevelopment projects and are therefore subject to changes that give free rein to heritage battles (Baron and Roseau 2016). Indeed, these stations are often considered places of high symbolic value (Baron and Roseau 2016) by the inhabitants of their anchor city, causing mobilizations and conflicts between actors.

This aspect is illustrated more specifically by urbanist Étienne Riot (2016) in a work devoted to Grand-Central Station in New York. Inaugurated in 1913, over the years it has come to be a symbol of New York City. During the 1970s, the station was threatened with destruction to give way to a new urban project, which gave rise to a protest movement by various actors. This led to the abandonment of the project, with the building inaugurated in 1913 being maintained as it was in its original condition. For Riot (2016), it is a true "heritage turning point" that contributes to transforming the relationship between the station and the city. From a "material" relationship to a place of networking for the city and potential urban projects, the rapport with the station becomes an "intangible" relationship, where the station acquires the status of an "urban icon" (Ethington and Schwartz 2006, p. 12). It is characterized by the integration into its materiality of a "share of the urban experience, that is, its culture and its representations" (Ethington and Schwartz 2006, p. 12). The "heritage turning point" described by Riot has thus made it possible to bring a change of perspective to Grand-Central Station in New York, now considered as a site where a continuity is built between the identity of the place, the city and even the country.

More recent works have questioned the role played by stations in the dynamism of their anchor territories. They point out that beyond their functionality, stations are also places marked by a heritage that would be shared and revealed on a daily basis (Marie 2014). A local creation and a common good are captured here, contributing to "drawing the territories into a virtuous dynamic", as demonstrated by the organization of exhibitions within stations (Bowie 2014). Beyond this tension between heritage development logic and functional logic, other issues emerge, observing other transportation places.

10.3. The heritage development of roads: production of places and territories

Since the 1990s–2000s, roads have been the subject of a heritage interest that contributes to transforming the assembly of places and associated territories.

10.3.1. *Roads and their diverse forms of heritage development*

Several roads and "paths" are today listed as UNESCO World Heritage sites. This is the case, for example, with a section of the Silk Road. Covering 5,000 km, it connects Chang'an and Luoyang (the central capital of China during the Han and Tang dynasties) to the Zhetysu region in Central Asia. Established between the second century BCE and the first century CE, this corridor was widely used up until the 16th century. It has facilitated circulation and exchanges between many civilizations, whether commercial exchanges or intangible elements: religious beliefs, scientific knowledge, cultural and artistic practices notably. Various sites along the route of this corridor are also listed as UNESCO World Heritage sites, such as Buddhist cave temples and posthouses.

Since the 1990s, several roads worldwide have also been the subject of a heritage interest that does not fall under an institutional approach. This interest is linked to the structure of the roads: their relief and their works of art, such as the roads of the Vercors in France. But it is also motivated by the picturesque depiction of the places and sites they pass through (such as Norway's tourist routes), as well as the myths they convey. These roads, once identified by the places they served and made it possible to reach, have become the very object of the journey. Among the most iconic examples is Route 66 (USA), which connects Chicago to Santa Monica. Decommissioned in 1985, the road has been the subject of a new enthusiasm since the 1990s, reflected by a series of local and national initiatives to ensure its preservation and develop tourism. For many Americans, Route 66 retains a mythical character and is imbued with nostalgia, insofar as it echoes leaving on vacation, individuals acquiring an automobile, and drive-ins and diners (Courbières 2018).

Several works have been devoted to these heritage roads, notably Route 66. Part of these analyses have focused on the practices of this road by users. Thus, in his thesis devoted to the history and construction of the myth of Route 66, historian Peter Dedek (2002) focuses on the relationship between it and those who use it. A few years later, anthropologists Kellee Caton and Carla A. Santos (2007) analyzed the experiences of tourists having traveled Route 66. This research, which was in line with the works devoted to heritage tourism, showed that tourists' experiences of their journey along this route were all different, and that their desire to travel it was not solely guided by "nostalgia". More recently, geographers Adam Payne and Douglas Hurt (2015) worked on a "geography of memory" along Route 66. They analyze the way in which the various "heritage themes" are presented to tourists and highlight the selective nature of this presentation, which focuses on the daily lives of individuals and communities along the road since the 1930s, on roadside work and on the old section of the road.

Mountain roads established as heritage objects have also been the subject of several studies, as evidenced by those of geographer Xavier Bernier (2009), who focused specifically on mountain roads in Switzerland. The author analyzes the functions of mountain road heritage and proposes to seek to model the territorial processes linked to its heritage development.

In order to illustrate in more detail how the heritage construction process of a road and the associated territorial restructuring are characterized, we will focus on the example of France's RN7 for the rest of this section, drawing on recent history and geography works that explore the subject (Flonneau 2016; Musset et al. 2019; Ballot 2020).

10.3.2. *National Road 7: from Roman road to a heritage object*

Linking Paris to Menton (Alpes-Maritimes department) over 996 km, the initial route of what would become the RN7 appeared as early as the first century BCE, in Roman times. Successively the path taken by the Romans to penetrate Gaul, the path of the kingdom of France and the imperial road, the RN7 has long been one of the country's main axes of circulation, whether at a local scale for its residents, or at a national scale for tourists on vacation, like other French national roads such as the RN6 and RN10. More specifically, it was the streams of summer traffic that gradually saw the RN7 enter into history.

The adoption of the French law on paid vacations in 1936 led several thousand motorists every summer from 1937 to take sections of the RN7 to reach their French Riviera vacation destination. The work/paid-leave pairing thus has a spatial counterpart in the Paris/French Riviera tandem, whose functioning is conditioned by the automobile and the road (Sirinelli and Rioux 2002).

The television reports of the time illustrate the long queues of cars loaded with luggage and make it possible to capture keys of the booming consumer society: the advent of individual automobiles for the middle classes, the liberation of bodies, summer vacations becoming commonplace, etc. Far from being a simple circulation route, the RN7 embodies the vector and emblematic figure of this period. The commissioning of the highway in the early 1970s would then contribute to draining an increasing part of the flows previously present on France's national roads (Flonneau 2016). This is a phenomenon observed particularly in the summer, when tourists now use the highway rather than national roads to reach their vacation destination on the French Riviera. From the 1990s onwards, a heritage interest in the RN7 emerged, reflected in different ways.

10.3.3. *The heritage developments of National Road 7: between mediatization and local initiatives*

Heritage interest in the RN7 initially emerged on a local scale, with the creation in 1990 of the first association dedicated to the RN7: *Mémoire de la Nationale 7*. Created by long-time residents of the municipality of Piolenc in France's Vaucluse department, the association first aimed to enhance everything related to road heritage.

The association's first temporary exhibitions brought together vehicles used in the 1950s and 1960s as well as objects, old Michelin road signs and markers, and old petrol pumps. A permanent exhibition was inaugurated in 2003, then again in 2012 in the form of a museum named *Mémoire de la Nationale 7*.

Several years after the creation of the association, a first book devoted to the history of the RN7 was published in 2003. Entitled *La Route Paris-Cote d'Azur*, it is the result of work carried out over several years by illustrator Thierry Dubois, who during a trip along the RN7 in 1995 was impressed by various aspects that mark the route: landscapes, buildings along the roadside, historical monuments, etc. The author retraces the history of the RN7 from Roman times to the 1990s, then presents a set of aspects, from Paris to Menton, that in his view are of a heritage character for the RN7.

La Route Paris-Cote d'Azur would spearhead a series of books from 2008 onwards. Written by road history enthusiasts or fans of vintage car rallies, they also trace the history of the RN7 since Roman times and mention all the emblematic places and heritage aspects that mark its route from Paris to Menton (Hardivilliers and Raffard 2008; Naddeo and Chabres 2010; Pagneux 2016).

The publication of these books occurred in parallel to an increase in the number of press articles and television reports appearing on the subject. Published in regional daily newspapers such as *Le Dauphiné Libéré* or magazines devoted to automobiles, the articles focus specifically on a theme addressed in the books or return to a fact or anecdote involving the road in one of the French departments or indeed one of the municipalities through which it passes.

Many video reports are also produced and disseminated in information magazines, discovery magazines and those devoted to automobiles. Lasting anything from a few minutes to 2 h, they focus partly on a particular aspect of the road: the restoration of a building or anecdotes relating to one of its sections. They also look at the road as a whole, proposing to follow one or several motorists in vintage vehicles. While evoking the history of the RN7, drivers also make stops at stages deemed key, explaining why, in their view, these milestones deserve to be dwelled upon more specifically, for example, because of the historical importance of the site, like the *Relais Les 200 Bornes* in Pouilly-sur-Loire in France's Nièvre department, or the castle of Lapalisse in the Allier department.

In parallel with these publications and video reports, actions are anchored directly on the RN7. Along the road, it is possible, first, to frequently come across objects and an iconography that refer to them, as shown in Figure 10.2. This fresco on the ground, situated in Lapalisse in the Allier department, whose color is a reference to the "Blue Road" nickname sometimes attributed to the RN7, represents the road's route from Paris to Menton. It thus depicts the historical character of the infrastructure.

Figure 10.2. *Example of a reference to National Road 7: the fresco on the ground represents the route of the road from Paris to Menton (Lapalisse, Allier) (source: Ballot (2017)). For a color version of this figure, see www.iste.co.uk/fretigny/transportation.zip.*

Since the mid-2000s, operations to valorize and restore artifacts along the route have also been undertaken by residents and local elected officials. These actors restore the road signage of the 1950s and 1960s, such as milestones and Michelin signs as well as municipality entrance or exit signs. They rehabilitate roadside establishments where motorists made stops during their journeys in the 1950s and 1960s: petrol stations, garages and truck stops. Lastly, they preserve and enhance the advertising murals, present on the facades of buildings located along the road and made to be as visible as possible to passing motorists.

These heritage actions are also deployed in the form of events and shows of various kinds. Organized annually by people passionate about the history of roads or those who enjoy outings in classic cars, rallies invite participants to spend a few days or even a week traveling the entire road or one of its sections. By traveling aboard vintage vehicles, the goal is to plunge back into the 1950s and 1960s and reconnect with the travel conditions of the time, making the journey no longer simply a matter of changing location, but a space-time already likened to being on vacation.

Since 2006, heritage actions have been introduced concerning various territories, notably municipal, crossed by the RN7. Driven by elected officials, tourism professionals or residents, they first took the form of associations, which generally aim to enhance and promote the heritage aspects of certain sections of the RN7. These actions also result in events local to the municipality, which are held on an ad hoc or permanent basis. They commemorate the highlights of the RN7's past, for example, by recreating traffic jams or a parade of vintage cars for 1 day. The municipalities are also home to several museums, such as the *Musée Mémoire de la Nationale 7* in Piolenc in the Vaucluse department, and temporary exhibitions are regularly organized.

The valorization of a route like the RN7 therefore takes two main forms, one of media coverage, often on a large scale, and the other of initiatives anchored on a smaller scale. The more specific example of municipality-based events makes it possible to understand the territorial restructuring triggered by this process of heritage development of the road (Ballot 2020).

10.3.4. *The road as a local-scale event*

Currently, four events are held annually or biannually along the RN7: *Faites de la Nationale 7* (Pougues-les-Eaux, in the Nièvre), *Le Bouchon* in Lapalisse in the Allier, *N7 en Fête* (Loriol-sur-Drome) and *Le Bouchon* in Tourves (in the Var).

The events *Le Bouchon* of Lapalisse and *Le Bouchon* in Tourves propose reenacting throughout the course of 1 day the municipality's traffic jams caused by departures on vacation in the 1950s and 1960s. Around 100 vintage cars, along with other classic vehicles such as mopeds, are invited to recreate the ambiance.

The *Faites de la Nationale 7* and *N7 en Fête* events each offer a parade and an exhibition of vintage cars and activities similar to those of the *Bouchon de Lapalisse*, along with a 1960s-style rock'n'roll concert organized as part of *N7 en Fête*. But other initiatives are expanding these events, such as the stands of *Faites de la Nationale 7* exhibiting vintage objects and materials used during road trips.

Since their establishment, these municipality events have proved a success that continues to increase in popularity, as evidenced by the constant increase in the number of participants year on year, which has allowed municipalities to become aware of the attractive and unifying potential of the RN7 with the public.

The scale of these events' reach has steadily increased. As a result, they are now known throughout France, and also in other European countries, such as Belgium and the Netherlands. This has led to them becoming perennial, reinforcing the mobilization of the RN7 by municipalities as an element of tourist attractiveness and leisure activities in the long term.

The actors of these territories are thus trying to reap the benefit of a transportation infrastructure, but following modalities related more to its symbolism rather than its functionality. The hypothesis that can be formulated is that the road, by becoming an object endowed with meaning and signification for these territories, gives rise to a new form of configuration of the links between network and territory. While network and territory are generally considered antagonistic in the organization of space, as they relate to different distance managements, also called metrics (Lévy 1993), it is rather their co-construction that renders visible the relationship between road and territory.

10.3.4.1. *Pougues-les-Eaux: National Road 7 as a vector for identity construction*

A city famed for its thermal baths from the Middle Ages until their closure in 1971, Pougues-les-Eaux represented an important stop on the RN7, also because of the presence of its casino, which attracted many Parisians and other travelers. With the demise of thermal baths, the municipality experienced a decline in its tourism activity, the thermal heritage (park and buildings) not being subject to specific upgrading measures.

Wishing to create an event bringing the municipality together, the tourist office of Pougues-les-Eaux decided in 2012 to set up a 1-day event, *Faites de la Nationale 7*. Conscious of the success generated by the RN7 at national level, the organizers saw a theme "that appeals" and the road as an "angle to exploit". Although other activities take place in Pougues-les-Eaux, such as *Jardipougues* (Festival of plants and gardens), the purpose of an event based on the RN7 was to attract a larger audience, offering an opportunity to discover the municipality during the event.

Thus, according to its organizers, while the RN7 offers a "common thread" for Pougues-les-Eaux, it is above all what gives the town its "identity". While partly through the existence of the event, it is above all the presence of the road itself that has seen the municipality come to be well known and seek to stand out from other territories, with its identification now occurring in varied respects.

Indeed, in order to remind tourists that Pougues-les-Eaux is indissociable from the RN7, a signage project for this "National 7 territory" is in the pipeline. It is also intended to make the municipality more attractive for more than simply the duration of the event, although more tourists have visited the municipality in recent years. The presence of the RN7 also brings symbolic value to buildings located beside a mythical road. Whether it is during the event or throughout the course of the year, they benefit from increased publicity, such as for the casino of Pougues. This proximity to the roadway seems to be a necessary condition: the organizers of *Faites de la Nationale 7*, for example, already considered offering activities around the thermal park, but its position 350 m away from the RN7 ultimately led to the project being abandoned.

10.3.4.2. *Lapalisse: National Road 7 as a vector for discovering the territory*

In celebration of the commissioning of the Lapalisse diversion in 2006, elected representatives of the Pays de Lapalisse community of municipalities (2006) decided to reenact a municipal-wide traffic jam. The aim was then to commemorate for "one last time" the significant automobile traffic experienced by Lapalisse throughout the 20th century due to the presence of the RN7.

This activity proved to be an "unexpected" success, reflected in the participation of 200 vintage vehicles and several hundred visitors. Elected officials and employees of the intercommunality have therefore decided to renew it on a biennial basis. Faced with the ever-increasing number of participants from one event to the next (12,500 in 2014, and 20,000 in 2016, according to the event's organizing committee) and the associated economic benefits, notably related to shop purchases, a "National 7 tourism" emerged.

Unlike Pougues-les-Eaux, where the road contributes to its identity building, in Lapalisse it is the territory that, for the organizers of the event, contributes to giving the road an identity. Indeed, for them, the attractions of the road differ according to the singularities of the territories passed through. In Lapalisse, it is characterized by the presence of the castle of La Palice, which, in recent years, has been the subject of enhanced development. This is reflected in the holding, since 2016, of a summer sound and light show in the castle courtyard, as well as by more visitors throughout the year, most often during vintage car rallies.

In this process of heritage development of the RN7, the links between the road and the territories are characterized by a consubstantial relationship. The road brings a specific meaning to territories and maintains an almost inseparable link with them to acquire a meaning in itself. For other transportation places, it is a new form of heritage development that comes into play: the so-called heritage development "of the present".

10.4. Transportation and heritage development of the present: the example of airports

For several years, discussions have been reflecting on the heritage development of airports, described here as heritage development "of the present", insofar as it applies to places "still in operation and subject to multiple uses" (Colin 2013, p. 78, our translation). These infrastructures are subject to rapid functional evolutions that operate over a short period of time (Adey 2006a), which raises a set of issues.

10.4.1. *The challenging heritage development of highly evolving places*

The heritage development process "applies both to an object and to an ideal reality [...], [in a logic of] recognition and development of buildings, inherited spaces, objects and practices" (Fagnoni 2011, our translation). This process therefore generally concerns objects that have lost their functional usage, or whose function finds a different meaning at present. While the stations and ports mentioned in section 10.2 are partly in line with this logic, the routes mentioned in section 10.3 constitute a specific case study, since most of them (including the RN7) still retain functional dimensions at present. The same can be said of airports.

Various social science works carried out on airports have questioned the place of the airport as a material and symbolic gateway signifying their anchor territory by its monumentality (Adey 2006b; Roseau 2008; Frétigny 2013). This process suggests that airports are established as heritage objects and identified with the city or even the country in which they are located (Frétigny 2013). But their constant evolution makes these operations difficult (Roseau 2012), with the question also arising of whether airports can retain their identity over time. The heritage development of airports leads to a "heritage shift" and to a particular form of heritage development: that is known as "of the present".

10.4.2. *An emerging heritage development*

Airports have always historically been linked to cities, but have not always necessarily been geographically close. Although initially airports specialized solely in air transportation, they have gradually been won over by urbanization, both literally, by being gradually incorporated into the urban fabric, and by the diversification of the activities they host (Drevet-Démettre 2015).

Thus, the change in scale of urban regions in recent years has led to placing the airport as an emerging urban centrality, located in the urban outskirts, and then increasingly solidly established. Where heritage is often traditionally associated with centrality, which is itself linked to power, airports thus become places of expression of power through monumentality. They are thus erected as synapses and gateways to metropolitan spaces.

The recognition of the airport, the terminal in particular, as a heritage object is a legitimization process that can be likened to that applied more broadly to the architecture of the 20th century (Pearman 2005). If heritage development is a process that occurs over a long period of time because of the differences of opinion it raises (Gravari-Barbas and Renard 2010), Ageron (2020, pp. 16–17) believes that, "the social acceptance of the airport terminal as heritage seems to be underway" (our translation). It is thus, to his mind, a "heritage shift" that applies in the relationship to the terminal, linked to its iconic character and its role as a metropolitan gateway. Passengers are faced with a "hybrid heritage development experience", divided between a "functional uniformization" on a global scale and a "distinctive identity that has an advertising purpose" and that is more broadly economic (Ageron 2020).

This experience is reflected in the staging, by the airport operator, of elements linked to the territories within which the airports are established. This is the case at Incheon International Airport (near Seoul), where an indoor garden was reconstructed from iconic elements of Korean gardens and landscapes, such as hibiscus, campanula, red pine and bonsai (Figure 10.3). While recalling local singularities, this development of the place also participates in a more generic equipment logic (indoor or outdoor) tending to spread in various major airports, whether the butterfly garden or the orchid garden and koi pond of Singapore-Changi, the "seven species" garden of Ben-Gurion-Tel-Aviv (Weiss 2010), or the gardens of Bangalore and Suvarnabhumi-Bangkok.

Figure 10.3. *The indoor garden of Incheon International Airport in South Korea (source: Frétigny (2023)). For a color version of this figure, see www.iste.co.uk/fretigny/transportation.zip*

Another "heritage shift" is that it is not only a matter of heritage as a conservation or revelation of the past but also a transplantation extoling modernity. The art exhibited at the airport is in many cases contemporary art, often present in the various airport spaces, from the giant bronze bear statue to the lamp of Swiss artist Urs Fischer, in the middle of the duty-free area of Hamad International Airport in Doha, to the works selected by the San Francisco Arts Commission. These form part of an artistic program designed for the city's airport, with which various other airports in the United States are similarly equipped. The museum micro-spaces set up in various airports tend, however, to occasionally exhibit antique works of art, whether the small annex of the Rijksmuseum, which has been present since 2002 at Amsterdam-Schiphol Airport, which initiated these spaces, or the Roissy-Charles-de-Gaulle "satellite M" gallery since 2013, both located in close proximity to shops. These forms of airport heritage development, as in other transportation places, also contribute to commercial objectives. They are indeed closely associated with a consumer experience, linked to the commercial activities of the terminal and the attractiveness of the airport for transiting passengers, who might be tempted to opt for another hub.

Figure 10.4. *The TWA Hotel at John-F.-Kennedy Airport in New York (source: Rhododentrites (2020), CC license BY-SA 4.0)[2]. For a color version of this figure, see www.iste.co.uk/fretigny/transportation.zip*

The difficulty of air transportation to compose with history is illustrated by the thorny heritage development of its infrastructure. Associated with a strong imagination of modernity and contemporaneity and with strong evolutions (notably the growth in the number of passengers), air transportation tends to strike its infrastructures with obsolescence. This makes their future particularly complex when they participate in still functional places, as Roseau (2012) showed with respect to the TWA terminal at John F. Kennedy International Airport in New York, which is of great architectural interest. Its most emblematic part was listed as a remarkable place by New York City in 1994 and listed on the National Register of Historic Places in 2005, protecting it from the risk of demolition. The terminal was designed by Eero Saarinen with a wing-shaped roof, and inaugurated in 1962. Its closure in 2001 gave way to a long period of uncertainty and controversy as to its future. It was not until 2017 that it was reinvested, reconverted, with new neighboring constructions, into a TWA Hotel, which reintegrates this heritage into a transportation logic (Figure 10.4).

It is also useful to question missing or concealed heritage developments in international airports, particularly in the area of migration (Frétigny 2017), in contrast to sites such as Ellis Island for maritime mobilities, which play a key role in the individual and collective migration histories of yesterday and today. Thus, airports enable a multiplication of forms of heritage resources, whether present or absent, to be called into question.

2 Available at: https://commons.wikimedia.org/wiki/File:TWA_Hotel_(87640p).jpg.

10.5. Conclusion

In this chapter, we examine several types of transportation places, which are the subject of different forms of heritage development. This choice of approaching them successively along with their heritage development process have allowed us to re-examine the so-called traditional heritage development processes and to highlight two main aspects.

First of all, heritage development contributes to deconstructing the theory according to which transportation places are "non-places", that is, these "functional spaces born of globalization, standardized and dehumanized, carrying a break with 'anthropological' places like the home" (Augé 1992, p. 100, our translation). While anthropologist Marc Augé, who is behind this term, identifies supermarkets or large hotel chains as "non-places", he considers above all that infrastructures and transportation places (airports, expressways, trains and airplanes) are the most representative, due to the anonymity characterizing each individual there, and the supposed absence of links and interactions between them. However, the heritage interest held with respect to transportation places and the more specific example of stations and airports tend to prove the opposite.

Second, the chapter highlights the inseparable link maintained between transportation places and the territories in which they are established and where they are no longer viewed solely based on their functional and service dimensions. Consequently, drawing on the example of the road, we can consider that new forms of territoriality are emerging, within which transportation places and territories intersect consubstantially.

10.6. References

Adey, P. (2006a). Airports and air-mindedness: Spacing, timing and using the Liverpool Airport, 1929-1939. *Social and & Cultural Geography*, 7(3), 343–363.

Adey, P. (2006b). If mobility is everything then it is nothing: Towards a relational politics of (im)mobilities. *Mobilities*, 1(1), 75–94.

Ageron, P. (2020). L'aéroport, la patrimonialisation paradoxale d'un lieu générique [Online]. Available at: https://www.aphg.fr/L-ae%CC%81roport-la-patrimo-nialisation-paradoxale-d-un-lieu-ge%CC%81ne%CC%81rique [Accessed 16 October 2023].

Andreu, P. and Roseau, N. (2014). *Paris CDG 1*. Editions B2, Paris.

Augé, M. (1992). *Non-lieux : introduction à une anthropologie de la surmodernité*. Le Seuil, Paris.

Ballot, P.-L. (2020). Identité de la route vs identité des territoires : formes et échelles de la patrimonialisation et de l'appropriation d'un axe fonctionnel national. Analyse appliquée à la route nationale 7. PhD Thesis, Université Grenoble Alpes, Grenoble.

Baron, N. and Roseau, N. (2016). Les gares au miroir de l'urbain. *Flux*, 103(104), 1–8.

Bavoux, J.-J., Beaucire, F., Chapelon, L., Zembri, P. (2005). *Géographie des transports*. Armand Colin, Paris.

Bernier, X. (2009). La patrimonialisation des routes : le cas de la Suisse. In *Mobilités contemporaines. Approches géoculturelles des transports*, Fumey, G., Varlet, J., Zembri, P. (eds). Ellipses, Paris.

Bowie, K. (2014). Patrimoine et ressources sensibles des gares. In *Quand la culture s'invite en gare. Art, patrimoine et créativité en partage*. Chaire Gare, Gares et connexions, SNCF, PUCA, Paris.

Caton, K. and Santos, C.A. (2007). Heritage tourism on Route 66: Deconstructing nostalgia. *Journal of Travel Research*, 45(4), 371–386.

Clot-Goudard, R. and Tillous, M. (2008). L'espace du réseau : du flux au territoire. Le tournant pragmatiste engagé par Isaac Joseph. *Tracés, revue de sciences humaines*, 15 [Online]. Available at: https://journals.openedition.org/traces/683.

Colin, C. (2013). Patrimoine du présent, fondements et limites : les équipements producteurs d'électricité dans les vallées de la Loire et du Rhône. PhD Thesis, Université Gustave Eiffel, Champs-sur-Marne.

Courbières, C. (2018). Dans le sillage du mythe : rémanence de la Route 66, *Communication et langages*, 1(195), 59–76.

Debarbieux, B. (1995). Le lieu, fragment et symbole du territoire. *Espaces et sociétés*, 82, 13–35.

Dedek, P. (2002). *Hip to the Trip: A Cultural History of Route 66*. University of New Mexico Press, Albuquerque.

Diener, A.C. and Batjav, B. (2019). Axial development in Mongolia: Intended and unintended effects of new roads. *Mobilities*, 14(6), 778–794.

Drevet-Démettre, L. (2015). Quand l'aéroport devient ville : géographie d'une infrastructure paradoxale. PhD Thesis, Université Bordeaux Montaigne, Bordeaux.

Dubois, T. (2003). *La route Paris - Côte-d'Azur. Petite histoire des Nationales 5, 6, 7 et de la Route Bleue entre Paris et Menton*. Drivers, Toulouse.

Duby, G. (1983). *Histoire de la France urbaine*. Le Seuil, Paris.

Ethington, P.J. and Schwartz, V.R. (2006). Introduction: An atlas of the urban icons project. *Urban History*, 33(1), 5–19.

Fagnoni, E. (2011). Reconversion territoriale : des dynamiques touristiques entre "devoir de mémoire" et "volonté d'oubli". In *Tourismes et territoires*, Bleton-Ruget, A., Commerçon, N., Lefort, I. (eds). Institut de recherche du Val de Saône-Maconnais, Saint-Just-la-Pendue.

Flonneau, M. (2016). *L'automobile au temps des Trente Glorieuses. Un rêve d'automobilisme*. Nouvelles Éditions Loubatière, Carbonne.

Frétigny, J.-B. (2013). Les mobilités à l'épreuve des aéroports : des espaces publics aux territorialités en réseau. Les cas de Paris Roissy-Charles-De-Gaulle, Amsterdam Schiphol, Francfort-sur-le-Main et Dubai International. PhD Thesis, Université Paris 1 Panthéon-Sorbonne, Paris.

Frétigny, J.-B. (2017). How are aeromobilities changing? Reviewing the literature on European airports. *Mobility in History*, 8(1), 123–132.

Gravari-Barbas, M. and Renard, C. (2010). Une patrimonialisation sans appropriation ? Le cas de l'architecture de la reconstruction du Havre. *Norois*, 4(217), 57–73.

Hannam, K., Sheller, M., Urry, J. (2006). Editorial: Mobilities, immobilities and moorings. *Mobilities*, 1(1), 1–22.

d'Hardivilliers, A. and Raffard, M. (2008). *Nationale 7. Un road-trip à la française*, Editions Transboréal, Paris.

Harvey, P. and Knox, H. (2015). *Roads: An Anthropology of Infrastructure and Expertise*. Cornell University Press, Ithaca.

Joseph, I. (1987). *L'enjeu-station. La station de métro comme espace public*. RATP, Paris.

Joseph, I. (1999). *Villes en gares*. Editions de l'Aube, La Tour d'Aigues.

Lazzarotti, O. (2003). Patrimoine. In *Dictionnaire de la géographie et de l'espace des sociétés*, Lévy, J. and Lussault, M. (eds). Belin, Paris.

Lévy, J. (1993). A-t-on encore (vraiment) besoin du territoire ? *Espace Temps, Les apories du territoire*, 51(1), 102–142.

Lombard, J. and Steck, B. (2004). Quand le transport est d'abord un lieu ! *Autrepart*, 32, 3–19.

Marie, J.-B. (2014). Gares, sous un même vocable, deux réalités. *Urbanisme*, 50, 59–60.

Meier, A., Ben Israel, A., Roded, B., Abu-Ajaj, I. (2016). Taming the road, tamed by the road: Sense of road as place among Indigenous Bedouin in an ethnic frontier in Israel. *Mobilities*, 14(2), 250–266.

Musset, A. (2018). *7 routes mythiques. Quand l'histoire se mêle à la légende*. Armand Colin, Paris.

Naddeo, J.-P. and Chabres, M.-S. (2010). *Eternelle Nationale 7. Au cœur de la France*. Editions Gründ, Paris.

Pagneux, D. (2016). *Redécouvrir la Nationale 7*. Ouest-France, Rennes.

Payne, A.A. and Hurt, D.A. (2015). Narratives of the mother road: Geographic themes along Route 66. *Geographical Review*, 105(3), 283–303.

Pearman, H. (2005). *Aéroports. Un siècle d'architecture*. Le Seuil, Paris.

Riot, E. (2016). Un tournant patrimonial: New York Grand Central (1903–1978). *Flux*, 103(104), 21–31.

Roseau, N. (2012). *Aerocity. Quand l'avion fait la ville*. Editions Parenthèses, Marseille.

Sirinelli, J.-F. and Rioux, J.-P. (2002). *La culture de masse en France : de la Belle époque à aujourd'hui*. Fayard, Paris.

Snead, J.E., Erickson, C.L., Darling, J.A. (2009). *Landscapes of Movement: Trails, Paths, and Roads in Anthropological Perspective*. University of Pennsylvania Museum of Archaeology and Anthropology, Philadelphia.

Weiss, E. (2010). Establishing roots at Israel's Ben Gourion Airport Garden: Landscapes of national identity. *Mobilities*, 12(2), 199–210.

List of Authors

Solène BAFFI
Géographie-cités
Université Paris 1 Panthéon-Sorbonne
Paris and Aubervilliers
France
and
Stellenbosch University
South Africa

Pierre-Louis BALLOT
CITERES
Université de Tours
France

Anne BRETAGNOLLE
Géographie-cités
Université Paris 1 Panthéon-Sorbonne
Paris and Aubervilliers
France

Nadine CATTAN
Géographie-cités
CNRS
Paris and Aubervilliers
France

Kaduna-Eve DEMAILLY
LADYSS
Université Paris-VIII-Vincennes-Saint-Denis
Saint-Denis
France

Clément DUSONG
LVMT
Université Gustave Eiffel
Marne-la-Vallée
France

Antoine FLEURY
Géographie-cités
CNRS
Paris and Aubervilliers
France

Jean-Baptiste FRÉTIGNY
PLACES
CY Cergy Paris Université
Cergy-Pontoise
France

Jean-Paul HUBERT
LVMT
Université Gustave Eiffel
Marne-la-Vallée
France

Christophe MIMEUR
MATRIS
CY Cergy Paris Université
Cergy-Pontoise
France

Damien MASSON
PLACES
CY Cergy Paris Université
Cergy-Pontoise
France

Mathilde PEDRO
Géographie-cités
Université Paris 1 Panthéon-Sorbonne
Paris and Aubervilliers
France

Juliette MAULAT
Géographie-cités
Université Paris 1 Panthéon-Sorbonne
Paris and Aubervilliers
France

Nicolas RAIMBAULT
ESO
Nantes Université
France

Index

A, B

C, D

E, F

N, O

P, R

S, T

U, V

W, Z

www.ingramcontent.com/pod-product-compliance
Lightning Source LLC
Chambersburg PA
CBHW070244190725
29593CB00003B/202